THE OIL SYNDROME AND AGRICULTURAL DEVELOPMENT

Lessons from Tabasco, Mexico

Sara Jeannette Scherr

PRAEGER

PRAEGER SPECIAL STUDIES • PRAEGER SCIENTIFIC

New York • Philadelphia • Eastbourne, UK
Toronto • Hong Kong • Tokyo • Sydney

Library of Congress Cataloging in Publication Data
Scherr, Sara J.
The oil syndrome and agricultural development.

Bibliography: p.
Includes index.
1. Agriculture--Economic aspects--Mexico--Tabasco (State) 2. Cacao--Economic aspects--Mexico--Tabasco (State) 3. Family farms--Mexico--Tabasco (State) 4. Petroleum industry and trade--Mexico--Tabasco (State) I. Title.
HD1795.T26S34 1985 338.1'0972'63 85-9509
ISBN 0-03-004617-3 (alk. paper)

To the memory of my mother,

Lois Helen Brown Scherr

Published in 1985 by Praeger Publishers
CBS Educational and Professional Publishing, a Division of CBS Inc.
521 Fifth Avenue, New York, NY 10175 USA

56789 052 987654321

Printed in the United States of America on acid-free paper

INTERNATIONAL OFFICES

Orders from outside the United States should be sent to the appropriate address listed below. Orders from areas not listed below should be placed through CBS International Publishing, 383 Madison Ave., New York, NY 10175 USA

Australia, New Zealand
Holt Saunders, Pty, Ltd., 9 Waltham St., Artarmon, N.S.W. 2064, Sydney, Australia

Canada
Holt, Rinehart & Winston of Canada, 55 Horner Ave., Toronto, Ontario, Canada M8Z 4X6

Europe, the Middle East, & Africa
Holt Saunders, Ltd., 1 St. Anne's Road, Eastbourne, East Sussex, England BN21 3UN

Japan
Holt Saunders, Ltd., Ichibancho Central Building, 22-1 Ichibancho, 3rd Floor, Chiyodaku, Tokyo, Japan

Hong Kong, Southeast Asia
Holt Saunders Asia, Ltd., 10 Fl, Intercontinental Plaza, 94 Granville Road, Tsim Sha Tsui East, Kowloon, Hong Kong

Manuscript submissions should be sent to the Editorial Director, Praeger Publishers, 521 Fifth Avenue, New York, NY 10175 USA

Acknowledgments

I welcome this opportunity to publicly acknowledge and thank the literally hundreds of people who inspired and assisted in this project. I hope even those not mentioned by name will understand how warmly I appreciated their help.

Those in Tabasco should receive first thanks. Former governor Leandro Rovirosa Wade; and Ing. Pedro Rodríguez Sierra, Lic. Aquiles Martínez del Pozo, Rosario Pérez Sastre, and many others in COPRODET and the Tabasco State government were generous to me both personally and professionally. Ing. Germán Velázquez of the Grijalva Commission and Ing. Contreras of the Promotora Chontalpa provided invaluable research assistance. Cacao Association and SARH Cacao Plan officials assisted generously with information, field trips, and advice on my study.

Mexicans are renowned for their hospitality, but nowhere more than in rural Tabasco. Over sixty farmers responded patiently, helpfully, and with friendly good humor to what must have seemed endless questions about their families and work. Few agricultural researchers can have chosen such a delightful project, where every farm visit brought a delicious chocolate drink or candy.

In Mexico City, Ing. José Rodriguez Vallejo and Ing. Adalberto Palma Gomez, long-time friends from my Papaloapan study, gave me useful advice and encouragement. Researchers from the Colegio de Mexico and Universidad Nacional de Chapingo willingly shared their findings and comments.

So many at Cornell University contributed to this work. To Professor Thomas T. Poleman I owe the inspiration for research on employment in agricultural development. As my thesis chairman, he encouraged intellectual breadth of interest and the flexibility so critical to research endeavors. Professor Davydd Greenwood was my guiding light in research methodology, and continually helped to sharpen my analytical skills. Professors Tom Davis, Daniel Sisler, Ken Robinson, and Robert McDowell provided invaluable input at all stages of research.

I was extremely lucky to be able to spend a year completing this project as a Visiting Scholar at the Food Research Institute of Stanford University. My thanks to those who made this possible, particularly Prof. Walter Falcon and Prof. Bruce Johnston, and to all the faculty and students whose work in Mexico, Nigeria, Indonesia, and Venezuela provided valuable insights into--and stimulating arguments over--agriculture and the oil syndrome. These include especially

Professors Scott Pearson, Clark Reynolds, Pan Yotopoulos, and Leon Mears; and students Robin Marsh, Paul Trapido, Bob McCleery, Paul Dorosh, and Steve Mink. John Lynam and Doug Perry, fellow FRI visitors for the year, provided the intellectual camaraderie I had not had since graduate school. I thank John particularly for his very helpful suggestions for my introductory chapters. Doug and Linda Perry also shared their home on my many treks from Modesto.

A special thank you for Margery Fletcher, who helped to prepare bibliography, index, and data displays, and did my editing and filing for the last year, all the while providing great company and political conversation. Thanks also to my old friend Lillian Thomas from Cornell, who also helped in editing and data displays; Joe Baldwin, who drew the maps and graphs; and Kay Byers, who typed my charts. For their assistance with computers, my gratitude to José Santos in Cárdenas, Tabasco; Prof. Bob Parsons at Cornell University; Dr. David Winsemius in Modesto, California; and Bob Wright of California State College, Stanislaus.

At Praeger, Barbara Leffel, Stephanie Grant, and Mia Crowley carried forward the work of publication, with enthusiasm, and much good advice for me. My sincere thanks for an enjoyable introduction to publishing.

I want to extend my very grateful thanks to the Tinker Foundation and Cornell University's International Agriculture Program for funding my Tabasco research; and to the Mellon Foundation for funding several months of research at FRI.

My sister Robin Scherr spent six months working with me in Tabasco. Her help was invaluable, and without the stimulation of her company, many ideas would never have developed. So many friends provided inspiration, company, and practical advice: Robin Marsh, Peter Ewell, Diane Wolf, Diana Callear, and Peter Berman especially.

Greatest thanks of all to the liberated men in my life. My father Dr. Merle S. Scherr first encouraged my interest in international affairs. My husband Dr. Alan W. Dappen interrupted his medical training to come with me to Tabasco, supported me through the writing of thesis and book, and has made a happy home with me despite our crazy schedules. My sons Benjamin and Nathan never quite understood "mommy's book about farmers," nor why they weren't allowed to play with the computer, but I thank them for adding so much joy to our lives during its production.

Preface

The role of primary commodity exports has long been one of the most important and hotly debated issues for economic analysis and policy in the developing nations. During the colonial period, primary exports were commonly associated with exploitation of producers. After independence, dependence on primary commodities--especially minerals and tropical plantation crops--continued, produced mainly by isolated enclaves. The economic development linkages hoped for from these export economies often failed to develop, and a large interpretive literature developed to explain this failure.

One of the more powerful arguments made against a primary export-dominated economy was that of Raul Prebisch and the Economic Commission on Latin America in the 1960s, who argued that primary commodities faced a structural deterioration in terms of trade. Yet the high international commodity prices of the mid-1970s led to another burst of optimism for the primary-export-led economy. Nowhere was this felt more strongly than in the oil-exporting developing countries, who saw their export earnings increase in some cases by a factor of 10.

But as with earlier commodity export booms, initial euphoria and optimistic development scenarios were dissipated by what appeared to be a whole series of phenomena related to the oil boom, which acted *against* development in the rest of the productive domestic economy--what is usually now called the "oil syndrome." The post-1979 decline in world oil demand and prices reminded producers that, as with any commodity, petroleum revenues are subject to large fluctuations. The instability of single commodity export dependence, and the short time frame within which petroleum resources would be available for export in most of these countries, led producing countries, and interested international observers, to become seriously concerned for the apparent deterioration in productive capacity of other sectors of the economy, especially manufacturing and agriculture, resulting from oil development.

For countries with high reserves of petroleum, and small populations (principally in the Middle East), these problems were less serious. Macroeconomic changes reflecting long-run dominance of oil exports approximated long-run conditions; specific sectoral developments could be accomplished through large subsidies. For the populous, agrarian-based developing countries with limited petroleum resources, however, the potential for deterioration of sectors on which the long-run health of the economy depended was extremely ominous.

There was particular concern for the agricultural sector, as governments watched with alarm the development of huge food import bills, accelerated rural-urban migration, and urban food riots. The prognosis for agricultural development was quite pessimistic, given rapidly rising non-farm wage rates, overvalued exchange rates constraining export production, and high inflation raising farm input costs.

Thus, when I first went to the tropical Gulf state of Tabasco, Mexico in 1979, where five years earlier one of the largest petroleum deposits in the world had been discovered, I fully expected to find an agricultural crisis stemming from the oil syndrome. Serious regional economic disruptions there were joined by macroeconomic policy for Mexico as a whole which was inimical to agricultural development. Instead I found a booming agricultural sector, largely smallholder-based, producing in many cases for export. The rural economy had successfully adjusted to conditions of high inflation and rapidly rising wages for rural workers, despite agricultural systems permitting little mechanization.

This book analyzes the factors responsible for the 1970s agricultural boom in Tabasco--the context and impact of national macroeconomic and agricultural policy, the development of Tabasco state resources and policies toward the rural sector, and most particularly, the dynamics of farmer and worker response to the transformations of the rural labor market resulting from petroleum sector expansion. The results, I think, are illuminating and thought-provoking. The most important finding was that stability and growth of Tabascan agriculture was due in large part to the adaptability and flexibility of "campesino" farm families who could integrate farm and off-farm employment, relying on little hired labor. Other farm groups did experience significant deterioration as a result of the "oil syndrome." A detailed analysis of the cacao sector is presented to illustrate the dynamics of farm production and farm labor allocation.

At the conclusion of my study, I was much more optimistic that agricultural sector disruption caused by the oil syndrome could in many circumstances be mitigated by specific policies. I spent much of the period 1983-85 comparing and contrasting Mexico's agricultural policy and performance with other large agrarian developing countries exporting petroleum. This research gave me both the context and confidence to present Tabasco's experience not as an anomaly, but as a source of useful lessons for agricultural policy in other oil-producing countries.

That an agricultural export boom could be found in Mexico, and in one of its highest wage regions, during this period is contrary to much "oil syndrome" analysis which concentrates mainly on macroeconomic phenomena. Mexico's macroeconomic performance was one of the most distorting of all the oil-exporters, yet its agricultural performance was better than than of most of the major developing oil-exporters except Indonesia. To understand why, one must go beyond the study of macroeconomic conditions, to examine specifics of agricultural policies and programs, and the structural characteristics of farming systems.

The outcome of agriculture-petroleum interaction is complex, depending on regional characteristics of demography, prices, farmer organization, and agricultural labor use, as well as economic policy at both the national and regional levels. Public policy formulation cannot rely on general models of the "oil syndrome;" careful--though not necessarily extended--field work to determine these characteristics will most often be required. I hope this case study of Tabasco, Mexico, will provide both a starting point for developing such research, and inspiration to policy-makers in "booming economies" alsewhere to seek and act upon the results.

Contents

List of Tables, Charts, and Maps

TABLES

CHARTS

Chart

Chart

MAPS

Introduction: "With Petroleum Wealth, To Sow Justice"

Through much of its history, the small state of Tabasco on the Gulf coast of Mexico has been isolated from the mainstream of national history and economics. A vast floodplain for the southern Sierra Madre, its life has ebbed and flowed with its surface waters. During the rainy season, half the state is under water.

The higher lands are fertile, and for many centuries have supported an agricultural population, who settled along the dikes of the many rivers. Outsiders--Mayans, Spaniards, Aztecs--sought tribute from the Tabascans, or trade, but wrought only minor changes in their way of life. The land--and their lives--were held fast by the roots of jungle and plantation trees, which alone could tolerate the swollen water resources. In Tabasco, basic sustenance came traditionally not from the tortilla, but from pozol, a liquid mixture of crushed maize and cacao beans.

But while ribbons and sheets of water still dominate the land above, below, Mexico's national oil company (Petroleos Mexicanos, S.A., or PEMEX) has discovered one of the largest deposits of oil and natural gas in the world. Since 1972, the entire region has undergone an extremely disruptive "petroleum boom."

Tabasco in 1980 was a place of astonishing contrasts and rapid change. Drawn by the magnet of petroleum development, men from northern Mexico, from Mexico City, from the Sierra had thronged south to Tabasco where they mingled conspicuously with the perplexed and unsettled native Chontal/Maya/mestizo population. The eminently agrarian nature of the Tabascan landscape and settlement pattern had radically altered with urban development, the erection of hundreds of oil wells and petrochemical processing plants, and the construction of highways and bridges, connecting what were previously islands of isolated settlements.

What has been the impact on rural life and on agriculture, the livelihood of most of the Tabascan people? The Zapata family represented thousands of other Tabascan farm families in their varied experience with the oil boom.

The Zapatas lived in a small ranchería halfway between Teapa, Tabasco's southernmost town, and Villahermosa, the capital. Those few miles traveled between city and farm, a quick trip on newly resurfaced roads, brought one to a new world. Could this tranquil land amidst the vestiges of rain forest really lie so close to boom-town Villahermosa and one of the world's great petroleum centers?

The Zapata home was a typical one, partly of wood, partly of the traditional palm, with thatched roof. It was large and airy, with two rooms and a separate kitchen--signs of relative prosperity by campesino standards.

The homestead actually included two other houses, occupied by families of Antonio and Rosa Zapata's married children. These were much smaller and simpler. The daughters-in-law did much of their household work, including cooking, with Doña Rosa.

The Zapatas were a prime example of the traditional Tabascan family farm, with its extended family, diversified production, subsistence security, and cash export crop. Elsewhere in Tabasco, many such farms had evolved to become either more fully commercial, or had been subdivided.

There were signs of new wealth everywhere--from Sra. Zapata's gas stove (which she used chiefly for boiling water, preferring the flavor of food cooked over wood) and sewing machine, to new household furniture.

All the grandchildren who were of age attended primary school. The range in education among the sons was extraordinary. The older two had finished only the fourth grade, back when their village of Quintana Roo was far more isolated. The third had finished primary school; the fourth finished high school; his younger brother was still in high school, but talked of seeking further schooling.

The family rose at 4:00 A.M. The rains had not yet begun in earnest, and most work had to be done before the mid-day heat. After a light meal of coffee and pozol, the family was off to work. Daniel accompanied his father to the cacao fields; Agustín left for his job as a health worker in Teapa; Jorge took the bus to temporary construction sites outside of Villahermosa, where he worked in petroleum company projects. In fact, more young men from the ranchería were off to the buses, than to the fields.

As Antonio and Daniel walked toward their cacao plantation, the morning was alive with vividly colored birds, including a flock of wild parrots. Already fishermen were out at the nearby lake. Fishing used to be a flourishing business here, but the water quality had gone down due to pollution from all the construction. Men could make a better living doing other things now, anyway.

They passed a large house obviously belonging to a wealthier rancher, and then some cacao plantations which were infested with weeds. Don Antonio shook his head to see the "cacaoito" in such bad shape. Their owner could not get men to work on his plantation. He wouldn't pay them enough, and they'd rather work for the petroleum company. The landowner could make more money building houses for rent in Villahermosa.

How different his own plantations! Huge trees--the "mothers of cacao"--shaded the cacao with their broad canopies. The ground below was nearly clear of weeds or debris, and the air felt clean and much cooler than the land beyond the plantation. Banana trees shaded the edges of the plantation. Though snakes were a hazard, this was a welcome place to work or rest. Rainbow-colored butterflies and moths flew here day and night.

The cacao trees themselves were perfectly shaped and full of shiny striped purple pods about a foot long. Daniel grinned at his dad, took out his machete to crack open a pod, and passed over a handful of cacao seeds so the two could suck off the sweet mucilage which covered them.

The men took out their new motor sprayer, a prized possession, and began to apply insecticide. They were careless with the chemicals, playfully spraying one another from time to time.

Taking a break for breakfast of pozol and fruit, Don Antonio and Daniel discussed digging a new drainage ditch on one side of the plantation. For this hard, hot work they planned to hire some experts and calculated the cost, which would run almost fifty percent above the minimum wage. Why didn't the brothers do this work?

They were too busy then in their other jobs. When harvest time came, however, and when they were in between other jobs, they put in a significant amount of work on the plantations. Antonio was rather proud of the fact that his sons helped him without even expecting a wage. Many of his neighbors paid their sons for such work (unheard of in the "old days"), and in many cases, in order to get the sons to stay on the farm, some of his friends had gone ahead and legally passed on their land while still alive.

On their walk back home, the two men laughed at a curious sight, which underlined the innate contradictions of this agrarian society undergoing rapid change: about fifty head of cattle were being herded up the road by a group of cowboys (complete with cowboy hats) waving out the windows of a Volkswagon "Bug"!

In addition to cacao, the Zapatas grew some maize, a variety of fruit trees, some beans and chiles. They also owned a number of small livestock and a few head of cattle. The family was no longer self-sufficient in maize and had to purchase quite a lot to feed several head of cattle. They did use their own maize for tortillas and pozol, however, for its taste was far superior to that which could be purchased.

They blamed the local shortage of high-quality maize in part on the fact that the traditional maize farmers in the foothills had abandoned their marginal farms entirely in order to work in the oil fields. Daughter-in-law Susana had earlier remarked that many had abandoned their families as well. Her friend worked in the office of the Municipal

President, who now received a steady stream of abandoned women seeking subsistence aid or help to prevent eviction from rented houses.

The house was full of babies and wonderful cooking smells when the men returned around 1:30. While they bathed, Rosa and her daughters-in-law completed a special barbequed chicken (slaughtered in honor of a guest to the community), seasoned with spices found in the woods nearby. These women did not have a very extensive kitchen garden but instead went on occasional hikes in the forest to find the items they wanted for cooking. They also minded the livestock, processed all the food produced on the farm, and helped with the cacao harvest. That day they had spent cooking lunch and cleaning the house immaculately; the next day they would host the new traveling rural health clinic in their home.

At lunch, Rosa, a hearty woman with "chutzpah," told stories about life in Quintana Roo when her children were young. Then, there were no doctors--she treated her boys herself, including doing minor surgery like separating a tied tongue with a butcher knife. She liked the new roads, which were being maintained well for the first time since she could remember. She herself rarely went to town. She complained of the "bad element" of men who now crowded the streets, disrespectful, and certainly responsible for the alarming increase in city vice and crime during the past few years. But a number of her neighbors earned extra money by making purchases in town to sell to village friends. Some even travelled to Veracruz or Mexico City, if they had relatives there, and brought back goods. Since prices were so much higher in Tabasco, they could make handsome profits on these trips.

Talk turned to the petroleum boom--cause of the high prices. Don Antonio had had no bad experiences so far, but he had heard other farmers complain that their fences were rotting away. Many were also blaming lower than usual cacao yields that year on the fumes from petroleum processing.

The family was doing well because cacao prices had been very high. Antonio was enthusiastic about the local Cacao Producer's Association, which he felt was responsible for the high prices. He had used nursery stock from the Cacao Plan program, and had begun to ferment some of his cacao beans at the local fermenter, which improved bean quality. Daniel, however, was suspicious that the Union was building fancy new office buildings in Mexico City, among other things, and figured that in fact they should be making even more on the cacao.

Afternoon chores included work on the new roof, tool repair, and harvest of some fruit growing near the house. But the family all relaxed in hammocks as evening arrived, and the breeze grew cooler and much more comfortable. The family stayed up later than they had last year, before the government extended electric lines to Quintana Roo.

It was not very dark outside anyway. All along the northern and western horizons was the glare of natural gas flaring. One drilling rig was outlined by the glow--it looked like nothing so much as a giant rocket ship. They were used to seeing it now. Rumor had it that the company would soon be drilling in their own ranchería. Don Antonio was a little concerned that the drilling would "poison" his land.

Would Don Antonio's sons be running the farm twenty years from now? asked their guest. "Oh, I think so. You can't go wrong with cacao. The young men may go away, sow their oats, make some money, but they'll be back." One of the older sons, back early from his construction job, nodded in agreement. But Agustín, who harbored strong ambitions to move permanently to town as soon as he could afford to do so, merely frowned as his mother threw him a knowing look.

Slapping at an odd mosquito, the family got into their hammocks, ready to sleep. Through the open window they could see the drilling rig outlined by the flare.

One wonders. Would the Zapata sons be there twenty years hence? Their lives had been changed by the petroleum boom in Tabasco, but not profoundly, and--more surprisingly--not for the worse. Could such campesinos survive in the modern industrial economy which Tabasco was likely to become?

It has rarely happened. Elsewhere in petroleum zones families, their incomes eroding, were breaking apart and abandoning their farms to become integrated at the margins of the new urban-industrial society. But such had not been the case for most farm families in Tabasco. The Zapatas felt content, secure, and optimistic about a future still solidly based on agriculture, regardless of significant participation in the petroleum economy. Moreover, it appeared that even the Mexican economic "crisis" of 1982 did not substantially threaten Tabasco's rural prosperity.

What was responsible for this unexpected and unprecedented flourishing of a peasant economy within a petroleum boom zone? Was it likely to endure? Was it relevant to other zones undergoing such rapid industrial transition?

When I first went to Tabasco, it was with the expectation that petroleum development would clash violently with the traditional agrarian society of that state. But in Tabasco the Mexican government simultaneously pushed for agricultural development in a big way--a response to the crisis in basic foods production and consumption in the 1970s. The Southeast, the "last agricultural frontier," was to play an important role in the country's rural reorientation. Furthermore, a strong and activist state government, funded by tax monies from PEMEX, coordinated a broad series of investments not only in agriculture, but in social infrastructure, tourism, fishing, and other sectors, throughout the state. Those responsible insisted that it *was* indeed possible to "sow justice" with petroleum wealth, as the governor had proclaimed in 1980.

Government policies to contain the negative effects of petroleum development, and to spread the benefits of the massive capital transfers more widely through the state, were only part of the explanation. A second was the state's exceeding good luck in having the petroleum boom coincide with a period of rapidly rising international prices for its major agricultural products. The negative impact on local inflation was far outweighed by the positive effect on internal distribution of income and political power.

Most important, the state's agricultural structure, based on small-scale, mixed commercial/subsistence family farms, proved highly favorable. The dominant "campesinado" showed more economic flexibility than "capitalist" type farmers, and greater stability in the face of falling relative farm prices. Family employment strategies permitted them to combine farm operation with employment in petroleum activities.

The marked mobility, levels of investment, response to price incentives, resolution of labor and wage constraints, and flexible interaction with the nonagricultural economy of the cacao farmers was impressive. The apparent multiplier effects of their prosperity--on rural employment, demand for consumer and investment goods, and on local and regional services--testify to the viability of rural-based development strategies. Their experience justifies a serious look at the potential positive role of the peasantry in petroleum-exporting economies.

Chapter 1 summarizes the key economic issues of agriculture-petroleum conflict and oil development in Mexico. Chapter 2 provides an overview of Tabascan economic history and the impact of the petroleum boom. Chapter 3 explains the cacao production system; producer organization, marketing, and technical programs; and documents the cacao boom of the 1970s.

Chapters 4-7 examine the production and employment characteristics of the four principal types of cacao producers, and their very different responses to Tabasco's petroleum economy. Chapter 8 looks at the employment characteristics of the petroleum economy, economic interactions between the different groups of cacao producers, and the resulting changes in the rural labor market in Tabasco. Chapter 9 concludes the book with a summary of the principal lessons of the Tabasco experience for agricultural development in a petroleum economy, and a challenge to rethink the role of mixed subsistence/commercial smallholders in areas of rapid economic change.

Devising a methodology to analyze agricultural response patterns to Tabasco's oil-dominated economy was challenging. Often, changes which first appeared due to petroleum development in fact were the results of internal dynamics, long-term state economic trends, or the dynamic agricultural growth of the 1970s. One thus needed to identify the relevant heterogeneity of physical and economic regions, producer resources, and institutions which might influence agricultural response. A methodology was designed to systematically link field studies of farm household strategies with regional and macropolicy models, and to investigate the *mechanics* of oil syndrome response as they operated in Tabasco.

So that the Tabasco story may be told without interruption, discussion of this methodological approach has been relegated to Appendix A, and notes on the primary data sources used in the study to Appendix B. For the development professional, however, these chapters provide essential background material.

Reference to many local terms and organizational names will be made throughout the study. A comprehensive glossary follows the Appendixes on page 293 for convenience to readers. It will be helpful to review it before beginning Chapter 1.

1

Agriculture and the Oil Syndrome

> *Estan sacando el petroleo y el gas; estan debilitando a las tierras para que no produzcan ya lo que producían antes de que vino Petroleos!*
>
> *[They are taking out the oil and the gas; they are weakening our lands so they no longer produce what they did before the petroleum company came!]*
>
> *--Mexican farmer, 1980*

THE OIL SYNDROME

The dramatic rise in world oil prices in 1973 set off a breakneck pace of economic growth in oil-exporting developing countries; Mexico was notable among them. But like the gold rushes of yesteryear, the boom was quickly over, and early optimistic hopes that oil wealth would lead to broadly based economic development in these countries were muted, if not dashed. Jahangir Amuzegar summarized the "mixed blessing" of oil wealth:

> *For most of the oil-rich countries, the tradition-shattering changes, the political uncertainties, the security risks, the economic bottlenecks, and the structural distortions are probably greater today than in the pre-1973 era* [Amuzegar 1982, p. 815].

He concluded his grim review by calling for improvement in government capacity for managing oil revenues, reordering of national priorities, and recognition that the level of oil revenues, and domestic monetary, fiscal, and exchange policies would be strongly influenced by developments over which they had no control (Amuzegar, pp. 834-35). As this book goes to print, the major developing oil producers are scrambling to readjust their economies and their public development policies after the severe disruptions of the early 1980s (Scherr 1984b; Croll 1984).

Resource booms based on rent-generation have long been the subject of economic analysis and a substantial body of literature that attempts to explain why huge inflows of resources and foreign exchange most often result in disappointing levels of over-all economic development. See, for example, Malakis and Reynolds' (1966) study of Chile's mineral export economy; Pearson's (1970) and Pearson and Cownie's (1974) studies on African oil and mineral exporters; and Nankani's (1979) review of mineral-exporting economies. Key explanations for these disappointing levels of development have been the lack of effective linkages between the boom sector and the general economy, problems of foreign exchange appreciation, expatriation of much rent-derived income, and/or a pattern of domestic rent distribution which does not contribute to economic development.

The experience of the 1970s oil boom resulted in a new body of literature on the "oil syndrome" (or "Dutch disease," as it is sometimes called, referring to the impact of North Sea oil development on the Dutch economy). "Oil syndrome" analysis emphasizes the distortion in relative rates of economic growth in various sectors of the economy (Corden and Neary 1982; Gillis 1983). Rapidly rising incomes in oil-exporting countries lead to increases in demand which cannot be so rapidly met by expanding supplies. In the case of internationally tradeable goods (manufactures and agricultural commodities particularly), part of the increased demand can be met through trade, which keeps domestic prices of these products lower than they would otherwise be. An appreciating exchange rate makes imports of these goods even more attractive and keeps their domestic prices even lower (Gelb 1984, p. 29).

By contrast, the increase in demand for services, construction, and other types of "nontradeables" leads to a major increase in their prices. As relative prices shift against tradeable sectors, the economy experiences a decline in commodity production, specifically manufactures and agriculture. This is exacerbated for the agricultural sector by a tendency for rising income per head to be associated with a shift in the composition of demand away from primary products.

The implications of "oil syndrome" analysis and the very poor performance of the agricultural and manufacturing sectors in almost all developing petroleum-exporters have led to justifiable alarm in the international community and inside the countries themselves. Ensuring food supplies is a particularly acute concern. Economic growth in middle-income oil-exporting countries has been accompanied by an enormous increase in demand for food, both basic staples and higher value foods, sometimes rising 6-7 percent per year during the boom years (Scherr 1984b).

To satisfy these levels of demand through domestic production is a challenge. Yates calculated that during the 1980s, food demand in Mexico would rise by 4.1-4.9 percent per year. In order for domestic production to supply this level of food demand, a two percent per year increase in yields would be required, in addition to placing 14-20 million hectares of arable land into production (Yates 1981, p. 40-41). Under oil syndrome conditions, such goals would appear impossible to meet.

The alternative is heavy dependence upon food imports. But sliding oil prices after 1981 reminded the oil producers that export commodity prices can fluctuate markedly. Dependence on food imports for too large a share of food supplies can be destabilizing during intermittent periods of low foreign exchange earnings and reduces the availability of foreign exchange for needed domestic investment and production.

A final consideration is the relatively short time frame within which most of these countries can expect to export large quantities of petroleum products. Economic development brought unexpectedly rapid increases in internal consumption of petroleum (World Bank 1980; Odell 1984); and proven reserves in many of the countries are projected to last only a decade or two (World Bank 1980; Mikesell 1984). Sectorally diversified economies--which in agrarian countries must certainly include a strong agricultural economy--are essential for stable economic growth. Yet, as several analysts have warned, "oil syndrome" declines in commodity-producing sectors, particularly agriculture, are not easily reversed as oil export revenues decline, owing to irreversible structural changes (Ul Haque 1982; Gelb 1984).

Devising agricultural policies suitable for a nation undergoing a petroleum boom--or even a less pronounced but substantial shift to a petroleum-dominated economy--is highly problematic. Efficient use of monetary and exchange rate policies to reverse antiagriculture bias is made extremely difficult by increasing international financial interdependence and unremitting pressures on the exchange rate so long as petroleum exports remain high (Warr 1984; Gelb 1984). With a relatively open economy, attempts to shift relative prices in favor of agriculture could require enormous subsidies. The huge differentials in productivity between agriculture and other sectors of the economy, at least in the medium-term, and the extreme mobility of agricultural labor, may convince policymakers that an ambitious push for broad-based agricultural development is unrealistic, and dependence on food imports is sensible.

Agricultural production in the actual regions of petroleum exploitation would seem particularly threatened. Labor scarcity may be acute during the period of intensive infrastructure construction; and local costs of production and consumption may rise at rates higher than the national average. Environmental damage from wells, pipeline projects, and construction; air pollution from flaring and chemical plants; and destruction of property either through expropriation or temporary eminent domain threaten farming. The political strength of the farming population in such regions tends to diminish, as oil interests become paramount.

Until the end of the 1970s, when economic constraints forced them to rethink their position, leaders in many nations had concluded that oil syndrome pressures on mass agriculture precluded successful development efforts. They attempted to meet skyrocketing urban food demand through imports and/or by setting up a small subsector of highly capitalized producers, in the hope that rapidly rising farm productivity would enable this subsector to compete for resources with nonfarm economic enterprises. In most cases, they could not, despite enormous public subsidies (Scherr 1984b).

*The International Record of Agriculture-Petroleum Conflict**

The experiences of Algeria, Iran, Nigeria, and Venezuela between 1973 and 1982 illustrate the widespread failure to achieve agricultural development in a petroleum economy. In Algeria, Iran, and Nigeria growth in production per capita was zero or negative throughout the period (see Chart 1.1),

---

*The reader interested in learning more about agricultural policy and performance in the countries discussed in this section should review the following sources. Venezuela: Negrón 1978; Quintero 1970; Kaplan 1980; Tugwell 1975; Trapido 1984; Martz 1980; Roseberry 1982; and Salazar-Carillo 1976. Algeria: Richards 1983; Braun 1982; Pfeiffer 1984; Megateli 1980; Schliephake 1977; Conway and Gelb 1984. Nigeria: Pearson 1970; Schatzl 1969; *Economist* 1982; Watts 1984; Watts and Lubeck 1983; Bienen 1983; Onoh 1983; Quinlan 1984; Gray 1978. Iran: Katouzian 1978; Madjd 1983; Looney 1977. Indonesia: Ooi 1982; Timmer, Falcon, and Pearson 1983; Glassburner 1984; World Bank 1980b; Papanek 1980; Mears 1984; Rose 1982. Comparative studies of economic development in the oil-exporting developing countries include: Roemer 1983; Gelb 1981; Gelb 1984; Gillis 1983; Ul Haque 1982; Timmer 1984; Nankani 1979. These include references to the experiences of Trinidad and Ecuador, as well as the countries discussed above.

Chart 1.1 Agricultural Production in Oil Exporting Countries

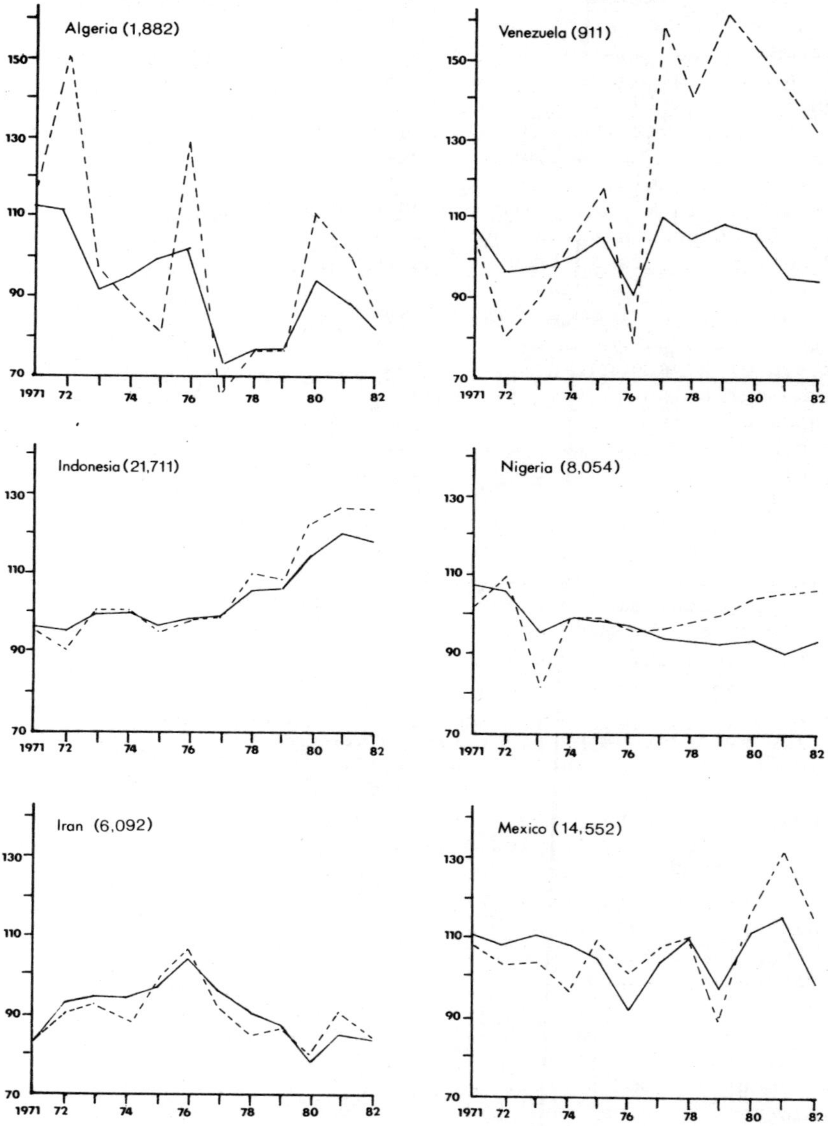

All Crops

Cereals

and rural-urban migration was extremely high. Venezuela's agricultural sector was already small and highly capitalized, the result of half a century of petroleum development. Its modest per capita growth in production was bought at an extremely high cost in government subsidies (Scherr 1985).

In all four countries, cereal imports in 1982 were far higher than in 1973, and the ratio in value of agricultural imports to exports rose steadily between 1973 and 1981 (Scherr 1985). Agricultural exports practically disappeared in Iran, Venezuela, and--except for citrus export to Europe --Algeria. In Nigeria, whose economy had grown rapidly in the 1960s based on agricultural exports, groundnut, palm oil, and cotton exports declined dramatically and cacao exports fluctuated on a downward trend.

This poor performance record seems attributable not only to the "oil syndrome" pressures described above, but to a pattern of government policies that exacerbated the problems of the agricultural sector. Iran, Nigeria, and Venezuela maintained fairly free trade in agricultural commodities through most of the period and followed urban-biased pricing policies to lower inflation by keeping food prices relatively low. Investment in agriculture represented a very low share of the total government budget, and of that, most was spent for subsidies of capital-intensive enclaves, or large-scale, capital-intensive infrastructure projects (Scherr 1985). Algeria, a centrally planned economy, managed thereby to avoid many of the macroeconomic distortions of the oil boom, but nonetheless followed a public investment program heavily biased toward urban, industrial activities (Conway and Gelb 1984).

The outstanding contrast to this gloomy picture was agricultural performance in Indonesia. Per capita food production in Indonesia showed a continual upward trend, led by major increases in staple rice production. While rice imports temporarily tripled between the early 1970s and late 1970s, by 1982 their level was the same as that in 1974. Agricultural exports outvalued imports in every year 1970-82 (Glassburner 1984; Mears 1984).

Despite monetary and fiscal policies which were hailed as unusually responsible--and several devaluations intended specifically to offset oil syndrome pressures--Indonesia's economy suffered from high inflation and an overvalued exchange rate through most of the period. But the Indonesian government pursued a program of smallholder agricultural development as a national priority, leading to much wider distribution of the benefits of oil development than was found elsewhere. Public policies--protectionist trade policies for agricultural goods, favorable pricing policies, and major investment programs to raise smallholder productivity --served not only to offset oil syndrome pressures, but to turn the internal terms of trade in favor of the smallholder agricultural sector.

Oil Syndrome and Agricultural Development: A Modified Analysis

The early evidence from cross-country comparative studies suggests that the sectoral shifts indicated in oil syndrome theory are not inevitable (Scherr 1984b; Scherr 1985; Gelb 1984; Ul Haque 1982). Gelb (1984, p. 29) held that the actual outcome would be affected by price controls, import policy, investment policy, traded sector investment, and price response, and the quality of economic investments made with oil revenues.

Ul Haque (1982) and Scherr (1985) emphasized the central role of government expenditure patterns in determining the relative prices of tradeables and nontradeables, given that most oil revenue is in the form of rent accruing directly to the government. In most other types of "boom" economies, returns are distributed far more widely among the population, with most of the returns going to producers (for example, farmers in an agricultural commodity boom).

The sudden shift in economic power to the state which occurs during a petroleum boom, combined with unprecedented levels of available revenues, has led to a fairly common pattern of public expenditures which contributes to agricultural decline. This pattern includes a preference for large-scale, capital-intensive development projects and for urban infrastructure and investment spending. Since government expenditures of this type require particularly heavy consumption of nontradeable goods (construction materials, transportation and communications services, skilled personnel), they may actually have as deleterious an effect on internal terms of trade between the tradeable and nontradeable sectors--and between the urban and rural sectors--as do direct oil syndrome distortions. Trade and domestic pricing policies which favor urban consumers, and the high wages paid for work in government projects, further distort terms of trade between commodity producers and the rest of the economy.

Thus, while macroeconomic policy to reduce oil syndrome pressures is certainly a key factor for agricultural development, of equal importance are sectoral spending policy and direct pricing decisions. The Indonesian case clearly shows that where agriculture is made a national priority for investment and development, and pricing decisions are made in favor of agricultural producers, petroleum-agriculture conflict can be minimized.

This is not simply a matter of throwing money at agriculture. A well-designed program for agricultural development must be predicated on microlevel understanding of petroleum-induced changes. Indeed, I believe that further development of "oil syndrome" theory will itself require more use of microlevel data.

For example, one of the more surprising aspects of agricultural development in the oil-exporting countries was the poor performance of the special agricultural enclaves that did receive very high subsidies and investments. These generally large-scale, capital-intensive farms were a marked contrast to the smallholders responsible for Indonesia's successful agricultural development. Why should smallholder development be more successful, or was this simply a fluke of the Indonesian economy?

The structure of Indonesia's economy was indeed very different from that of petroleum exporters such as Mexico. In 1980, 80 percent of the population was rural, as compared with only 33 percent in Mexico. The farm size distribution was quite different, more egalitarian, with smaller average farm size. Mexico's economy was far more industrialized and diversified; wages were far higher; its economy was integrated substantially with that of the United States; and the technology for increasing rainfed maize production was less promising than that which already existed for irrigated rice in Indonesia.

But I believe there are compelling microeconomic explanations for differing agricultural performance by different types of farm producers. I will argue in this book that given the conditions of a petroleum economy, even a country such as Mexico could benefit substantially from a smallholder-oriented agricultural policy. A major commitment of public revenues from oil for this purpose is not only justifiable, but essential to prevent irreversible agricultural sector decline and related problems of excessive agricultural imports, rural-urban migration, and unemployment. The case is even stronger for other nations in which the importance of the agricultural economy is greater--Nigeria, Cameroun, and Ecuador, for example.

The relative success of a smallholder strategy stems from the microeconomics of agricultural production and household decision-making strategies regarding the allocation of labor and other agricultural inputs. Because these factors are so critical to understanding the aggregate impact of petroleum development on agriculture, they need to be examined by macroeconomic policymakers, as well as agricultural production specialists. I thus ask the forebearance of those readers who may not usually be interested in farm management analyses to skim, at least, the chapters of this book that explain the microeconomic issues involved. The specific example of Tabasco, Mexico, will be used to illustrate and support the case for smallholder-oriented production policies in a petroleum boom economy.

The Case of Tabasco, Mexico

The dramatic recent petroleum discoveries and production gains in Mexico came almost entirely from three fields in the "Reforma" area in southern Tabasco and northern Chiapas states. Due to mountain barriers south of the Reforma field, most of the generated local economic activity in new drilling, exploration, and pipelines has been centered in Villahermosa, the state capital of Tabasco, and Cárdenas, thirty miles to the west.

This is the heartland of a densely settled agricultural region dominated by smallholder farmers. In fact, the last remaining agricultural frontier in Mexico--the Southeast, comprising the states of Tabasco, Veracruz, Chiapas, Campeche, and the Yucatan penisula--sits over a vast pool of exploitable oil and natural gas.

As one would expect, Tabasco, even more than the rest of Mexico, suffered extreme disruptions from the oil boom. But despite them, agricultural production in the state rose throughout the decade of the 1970s. Maize production declined, but cacao, bananas, coconut, and beef production reached unprecedented levels. Although there was extremely high participation by rural workers and farmers in petroleum-related employment, rural emigration rates were relatively low. This book attempts to explain this extraordinary contradiction, and analyze its implications for agricultural development policy.

Before going on to the Tabasco case, it would be helpful to review its context in the general Mexican oil boom. The section below summarizes the principal features of Mexican economic development in the 1970s, "oil syndrome" pressures on agriculture, and the fairly successful--by international standards of oil-exporting countries--agricultural production policy and response.

THE OIL SYNDROME IN MEXICO

Mexico had a long and eventful history of oil production well before the extraordinary discoveries of the past decade. Map 1.1 shows the major historical zones of production.

At the turn of the twentieth century, an American began exploration for oil on the east coast of Mexico near Tampico. His finds were the first in what would prove to be an oil-rich area known as the "Golden Lane." The major oil companies soon moved there, and by 1921 Mexico was the second largest oil producer in the world after the United States. These fields were ruined by overproduction, however, and later disputes between the companies and the Mexican government over labor policy in the mid-1930s led to Mexico's nationalization

MAP 1.1 Location of Mexican Hydrocarbon Production

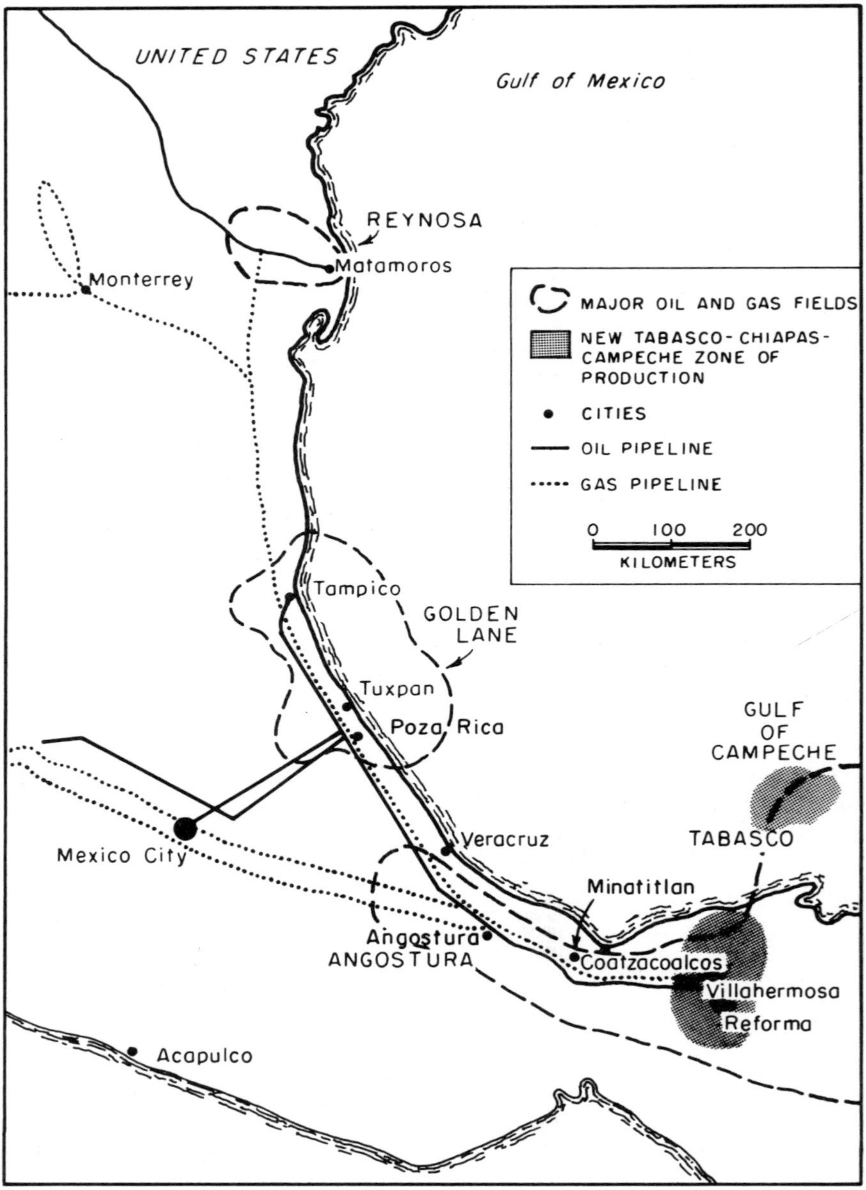

Source: Adapted from H. O'Shaughnessy, *Oil in Latin America* (The Financial Times Ltd.: London, 1976); and Mexico, Secretaría de Programación y Presupuesto, *La Industria Petrolera en Mexico*, pp. 353–55.

of production and the formation of Petroleos Mexicanos or PEMEX (Bermudez 1963; Bermudez 1976; Rippy 1972; Meyer 1978).

PEMEX in the next few decades discovered new fields around Poza Rica, Veracruz (the "New Golden Lane") and further south by Angostura, Veracruz. The twin cities of Minatitlán and Coatzacoalcos in southern Veracruz became an important Mexican petrochemical and distributing center. In the 1960s, petroleum and natural gas were found in Tabasco, in the municipios of Comalcalco, Cárdenas, and Macuspana.

Mexico in 1972 had by far the most expert and experienced petroleum industry personnel in the developing world. But attention to new reserves had slackened due to inadequate investment in exploration, and the industry was in a period of crisis (Bermudez 1976, p. 16).

The 1972 discovery of a huge, rich field in the Tabasco/Chiapas area took most of the country by surprise. A new technique of deep drilling revealed a giant Jurassic-Cretaceous "reef." Total production of crude rose from only .442 million (m.) barrels per day to 2.2 m. barrels per day in 1980 (PEMEX 1980). Estimated reserves of oil, gas equivalent, and gas liquids reached well over 200 billion (b.) barrels by 1982, a great deal more than Alaska's North Slope or the U.K.-Norwegian North Sea. Chart 1.2 shows changing production levels 1938-79.

Mexico's Economy During the Oil Boom

With official announcement of the oil discoveries in 1974, the pace of economic activity in Mexico accelerated. Rapid inflation, despite an average rate of growth in real income 1972-76 of only 4.8 percent, led in 1976 to devaluation of the peso for the first time in decades. For the three years 1978 through 1980, Mexico's GNP grew eight percent per year in real terms (El Mallakh et al. 1984, p. 10).

The economy was substantially altered during this latter half of the 1970s. Between 1975 and 1980, crude petroleum exports grew from 7.3 to 61.6 percent of total exports. The share of public investment allocated to the petroleum sector grew from 19.5 percent in 1976 to 29 percent in 1979 and increased further in 1980. The share of federal government income generated by oil revenues increased from 5 percent in 1976 to 29 percent in 1979.

Despite rapidly rising oil revenues, the discovery of oil and natural gas did not in fact lead to an immediate availability of substantial capital surpluses for general economic development. As Antonio Bermudez, former director of PEMEX in Mexico, forewarned, most revenues from oil and gas sales were ploughed back into exploration, drilling, petroleum infrastructure, and payment of petroleum development debt (Bermudez 1976, p. 25).

Chart 1.2 Petroleum Production in Mexico, 1938-79

(Crudes and natural gas)

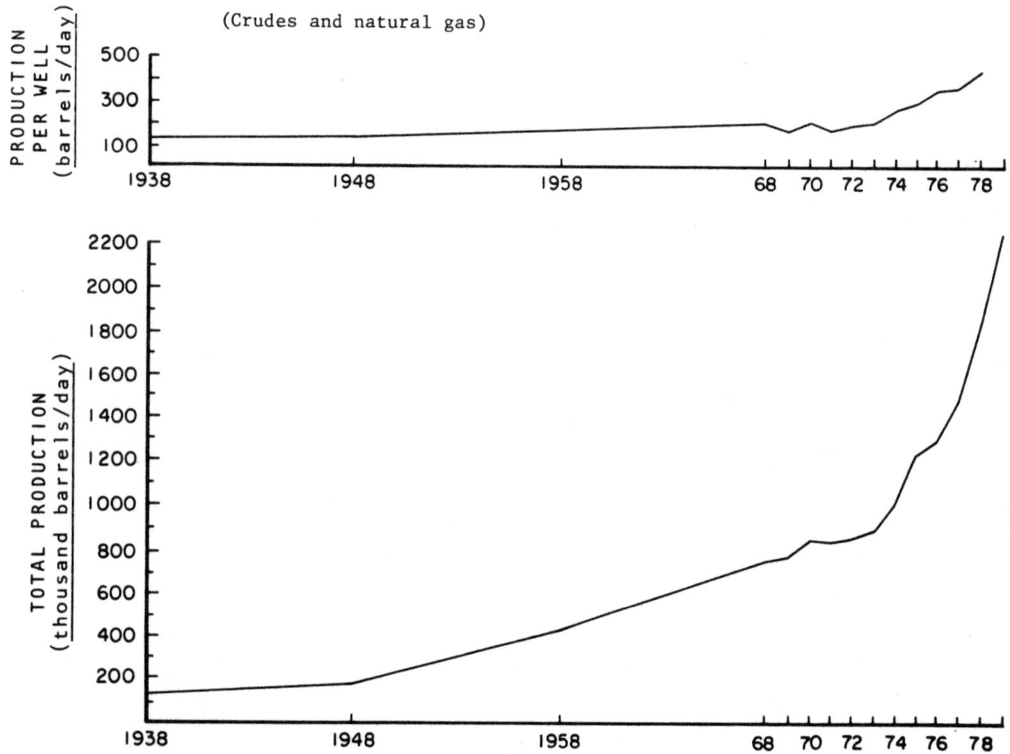

Source: Mexico, Secretaría de Programacion y Presupuesto, *La Industria Petrolera en Mexico*, October 1980, Chart II-32, p. 96.

Thus Mexico's official pledge to "sow the petroleum" during the 1976-82 sexenio--including major new programs for the agricultural sector--in fact meant dramatically increasing the debt load of the country (using future petroleum production as collateral) and inflationary expansion of the money supply to finance ambitious public projects. In April 1980, the ambitious Global Development Plan, and in May 1980, the National Plan for Agro-Industrial Development were announced.

By 1982, Mexico's public foreign debt was second only to that of Brazil--having more than tripled since 1975. Rapid growth in the money supply and government deficits raised the inflation rate to over 30 percent per year. In 1981, the economic system fell into crisis as these were joined by external factors. A temporary world oil glut occurred in 1981-82, pushing prices down, and Mexico had to reduce its selling price. U.S. interest rates rose sharply, increasing Mexico's debt burden. Speculation against the peso in the summer of 1981 resulted in an outflow of US$4 billion. Mexico was forced to officially devalue its currency in mid-1981; thereafter began a crawling depreciation of the peso through early 1982 (El Mallakh et al. 1984, pp. 44-46).

The devaluation, combined with reduced government spending and continued low oil prices, led to a severe economic contraction during 1982 and 1983. A sharp devaluation in March 1982 sent inflation soaring to over 60 percent later that year.

But by 1984, inflation was substantially reduced; economic growth was positive, if modest; and Mexico's short-term debt crisis (if not its long-term debt problems) had been resolved by the international banking community. From a position of serious overvaluation at the end of the 1970s, by 1983, the peso was probably undervalued (McCleery and Reynolds 1984). A major decline in imports substantially improved Mexico's foreign trade balance. Internal demand for petroleum, which had been growing very rapidly due both to high economic growth and to subsidized prices, declined after the government doubled the domestic price of petroleum in 1982. Gross industrial product in 1983 was only 89 percent of the 1981 level; by 1985 it was estimated to be nearly back to the 1981 level (BANAMEX 1984, p. 289).

Agricultural Performance During the Oil Boom

The potential effects of the "oil syndrome" were perceived by intellectuals and the development policy elite in Mexico at the very beginning of the oil boom (for example, Yuñez 1978). There was wide public support for a development policy which would avoid the problems of "petrolization" suffered by

Venezuela (*Foro* 1978). Official statements, from the Presidential Address in 1976 to the Energy Program of 1980, pledged that petroleum production would be held down to levels at which the general economy could efficiently absorb its revenues. But the pressures for more production, both internal and external, were overwhelming. Oil syndrome distortions were exacerbated by the further economic expansion induced by massive deficit spending and international borrowing (Székely 1983).

The oil syndrome was not in fact as serious in Mexico as in other oil-exporting developing countries. In part, this was because its economy was already highly diversified. In 1979, 23.8 percent of GDP came from manufacturing; 26.9 percent from commerce; 23.3 percent from services; 4.8 percent from petroleum and basic petrochemicals; 6.7 percent from construction; and 10.1 percent from agriculture, livestock, forestry, and fishing. Transport and communications, electricity, and mining together accounted for the remaining 5.5 percent of GDP (1.1 percent was subtracted from the total for banking services). (El Mallakh et al. 1984, Table 2-1.)

As a percentage of GDP, sectoral contribution changed very little between 1975 and 1979. The share of the agricultural sector declined by one percentage point, from 7.1 to 6.2 percent, while that of the livestock sector rose from 2.5 to 3.1 percent of GDP (El Mallakh et al., p. 12). This stability was critical, since agriculture's importance to the economy far exceeded what its share in GDP would indicate. Together, agriculture, livestock, and fisheries still employed one third of the economically active population of the country in 1979.

The Mexican population grew approximately 3.0 percent per year between 1970 and 1982. Wide year to year fluctuations in rainfall conditions,* and statistical problems with agricultural data in Mexico† make it difficult to analyze the trend in agricultural production during that period with

*The Ministry of Agriculture and Water Resources, SARH, reported that while average rainfall levels in every year from 1975 to 1980 were two to 12 percent below average, and in 1982, over 15 percent below average; in 1981, 1983, and 1984 rainfall was eight to ten percent above average levels (BANAMEX, p. 294).

†Yates provides an excellent summary and the interpretation of Mexican data sources on agricultural land use, production, yield, livestock numbers, irrigation districts, farm size, and land tenure (Yates 1981; Appendix).

confidence. Yates estimated that production during the 1970s rose somewhere between 2.5 and 3.5 percent per year. All figures show declining per capita crops production from the late 1960s to around 1976--the much publicized "agricultural crisis." Thereafter, the FAO index showed a modest trend of increasing per capita production until 1982; in cereals production, the increasing trend, particularly after 1979, was more marked (FAO 1982a). A major CESPA study showed the turnaround taking place after 1977 (SARH, CESPA 1982). An index of agricultural production used by BANAMEX showed an even steeper upward trend between 1976 and 1984 in agricultural production (BANAMEX 1984, p. 293). The World Bank reported that over-all agricultural production rose 3.4 percent per year 1970-82, or .4 percent per capita. Given the poor performance prior to 1976, this would indicate a substantially higher per capita growth rate in the latter part of the period.

The agricultural trade balance also appeared good relative to other oil-exporting developing countries, despite the overvalued peso. The volume of exports rose slightly 1970-80, with a major increase 1972-80 in the real value of exports. The average ratio of agricultural imports to exports for 1973-81 was 1.00, compared to 2.37 in Nigeria, 5.02 in Iran, 10.17 in Algeria, and 13.45 in Venezuela. Only Indonesia, with a ratio of .68, had a more vigorous export performance (FAO 1982b). The volume of Mexican exports of tomatoes, coffee, cacao, bee honey, garbanzo, and melons rose substantially. Exports of strawberries, sugar, and live cattle declined, not due to declining production but to rising domestic demand. Otherwise, the only declines in exports were cotton, which after 1977 began to rise again, and henequen. There were wide flucutations in beef exports (SARH, CESPA 1982, Volume IV). Even during the 1978-81 period, agricultural exports made up a fourth of Mexico's external trade, although only 3.4 percent of the harvested area was devoted to export crops in 1977. Production per unit land was worth three times that for as much labor (SARH, CESPA, Charts III-2, III-5).

There were large increases in the level of imports for milk, sorghum, and soy, as well as for modest quantities of lard, hides, and brown sugar. There were wide fluctuations in the level of imports for maize, wheat, barley, and cottonseed (SARH, CESPA 1982, Volume IV). Staple food imports tripled between 1968-72 and 1977-80, but by 1982, they had reverted to earlier levels--a result of declining demand during the recession, devaluation, and rising domestic production. The total value of agricultural exports was higher than imports in all years except 1980 and 1981 (BANAMEX, p. 299). Chart 1.3 shows fluctuations in agricultural trade for the five principal traded products.

Chart 1.3 Agricultural Trade in Mexico, 1965-80

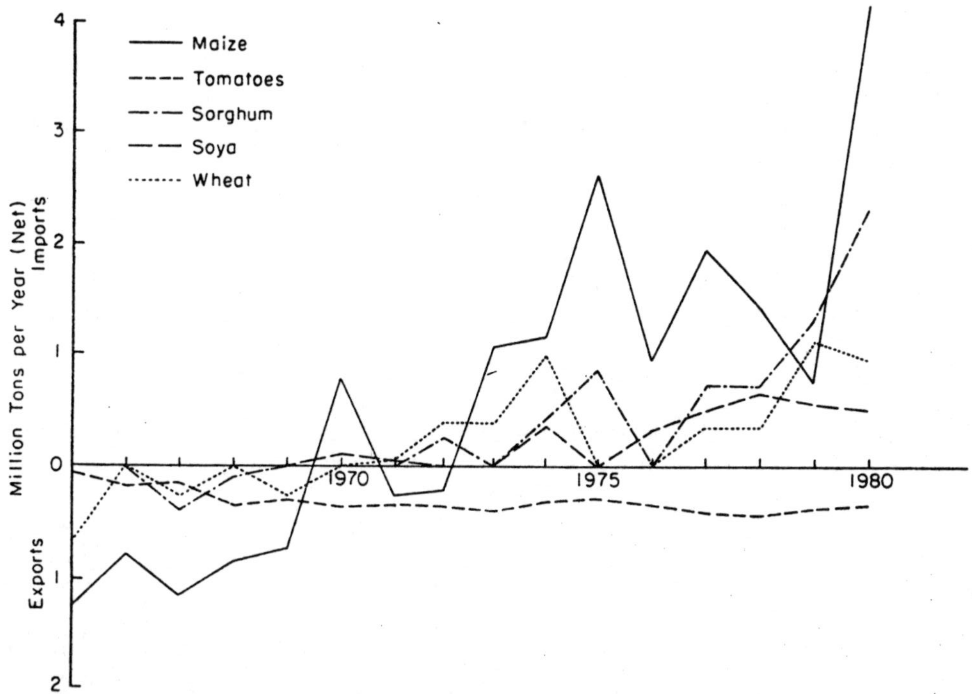

Source: Mexico, Secretaría de Agricultura y Recursos Hidráulicos, *Econotécnia Agrícola: Situación de la Agricultura Nacional en 1980 y Algunas Consideraciones Sobre el Mercado Internacional*, Vol. V, No. 1, January 1981.

Relative prices can largely explain the dramatic differences in production performance for different subsectors, shown in Cahrt 1.4. Although actual crop prices increased fivefold in deflated pesos 1970-80, the overall crop price index varied between 70 and 84 percent of the 1950 price levels, with highest real prices in 1973-78. BANAMEX figures showed a steady downward trend in deflated guaranteed prices between 1974 and 1981 (BANAMEX, p. 297). For livestock products, actual prices quadrupled, while deflated prices stayed at 80-81 percent of the 1950 levels, declining by 1980 to 71 percent of the 1950 level (Yates 1981, p. 214). There were no official price controls on meat products or fruits or vegetables, although there were occasional controls on meat and cattle exports.

Production of long-cycle fruits and forages rose extremely rapidly throughout the entire period, and vegetable production rose rapidly with the exception of a brief period 1974-76. Livestock production responded vigorously to rising internal demand based on rising consumer income during the oil boom. Even with impressive production gains during the 1970s, Mexico experienced shortages of beef.

By contrast, fiber production declined steadily during the 1970s. Basic foods, oilseeds, and industrial crops showed stable or slightly increasing production. Between 1969 and 1981, maize production hovered between eight and ten million tons per year, though demand rose from eight to 12 million tons over that period (Hall and Price 1981). At the same time, there was a decline in yields for maize and beans until the end of the decade (SARH, CESPA 1982, Vol. XIII). Given the stable or declining real maize prices over this period, as part of a policy of cheap urban food, the relative stability of maize production was a remarkable sign of the persistence of campesino producers. Larger commercial producers, and most of those with irrigated lands, switched to more remunerative products: the fruits, vegetables, and livestock products demanded by the increasingly prosperous middle classes (Rodríguez V. 1981).

There is thus substantial evidence that aggregate per capita agricultural production increased--in some sectors very rapidly--during the oil boom. Yet at the same time there was a huge apparent outflow of workers from the agricultural sector to other sectors of the economy. Between 1972 and 1982, the share of the EAP whose primary occupation was in agriculture declined from 45 to 34 percent--a greater proportion change than in Algeria, Indonesia, Iran, or Nigeria. Only in Venezuela (whose agricultural work force was already quite small) was the outflow from agricultural work more rapid (World Bank 1984).

Chart 1.4 Agricultural Production in Mexico, 1968-80

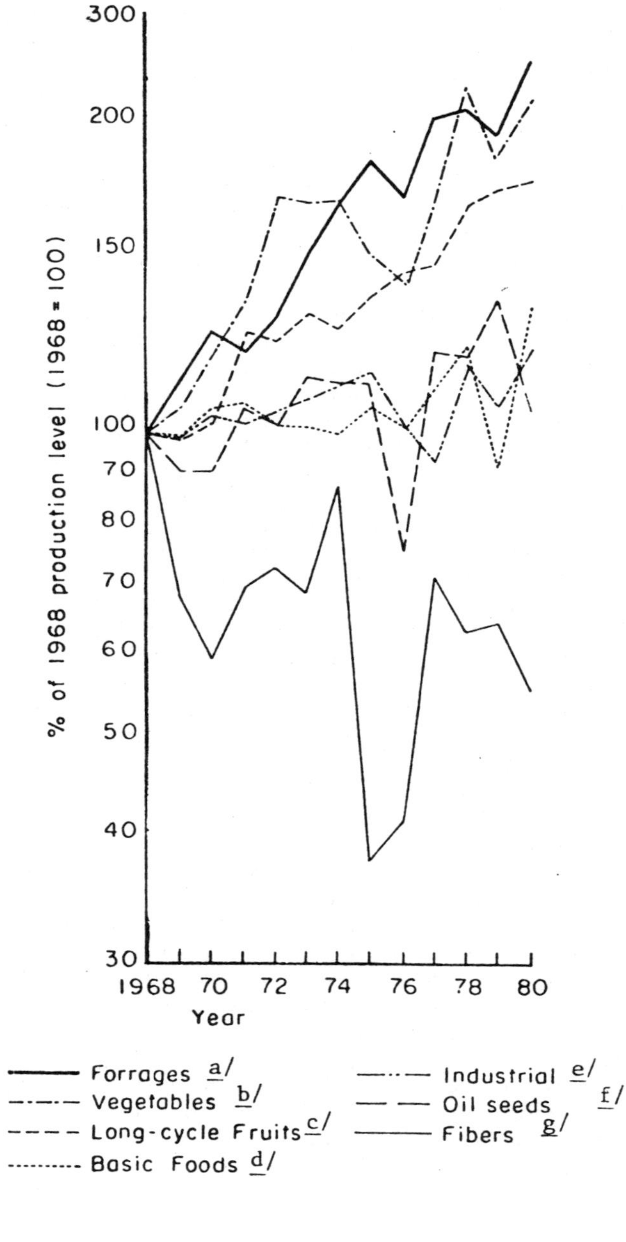

Chart 1.4 continued

Key:

———————	Forrages[a]
—.—.—.—.	Vegetables[b]
- - - - - - - -	Long-cycle fruits[c]
....	Basic foods[d]
—..—..—..	Industrial[e]
— — — —	Oil seeds[f]
——————— (thin)	Fibers[g]

Notes:

[a]Forrages include: alfalfa, beetroot, grain sorghum, canary seed, and grain and forrage chickpea.

[b]Vegetables include: garlic, beets, onion, peas, dry chile, green chile, corn, potatoes, tomatoes, green tomatoes, and sweet potato.

[c]Long-cycle fruits include: avocado, bitter cherry, almond, plum, country plum, coconut, apricot, peach, pomegranate, guayaba, fig, lime, lemon, mamey, mango, apples, pear, quince, oranges, nut trees, papaya, peron, bananas, tamarind, tejocote, grape, date, and grapefruit.

[d]Basic foods include: rice, beans, wheat, and maize.

[e]Industrial crops include: barley, cacao, coffee, sugar cane, tobacco, and vanilla.

[f]Oilseeds include: sesame, peanut, safflower, coconut, higuerrilla, linseed, soy, cottonseed, olive, and oil palm.

[g]Cotton is included in both fibers and oilseeds; coconut is split with 90 percent counted as an oil crop and 10 percent as long-cycle fruit.

Source: Mexico, SARH, Direccion General de Economía Agrícola, *Econotécnia Agrícola*, January 1981.

What exactly was happening? The few regional studies of Mexican petroleum centers during the oil boom show mixed results--declining maize production, high rural emigration, and rising production of high-value crops and livestock. Frauvergue (1980) described the steep agricultural decline of the Salina Cruz enclave around the refinery. Verduzco (1984) documented the outmigration from rural regions in which cattle production took over from campesino agriculture in Puebla. Pucciarelli (1979) described the agricultural changes around the Las Truchas steel complex in Michoacán, with high underutilization of irrigation district capacity, and a transition from basic foods production to high-value crops.

Zavala de Cossio (1981) described the enclave nature of development around Minatitlán and Coatzocolcos; Nolasco (1979) documented a sharp decline in agricultural employment in this region 1970-78, and workers' migration to "lost cities" around the major petroleum operations. Official statistics showed maize production in Minatitlán municipio declining from 25,000 hectares in 1971 to almost none in 1977 (SARH 1978). A major study undertaken by the state of Veracruz indicated that despite average per capita income in Poza Rica far above the state average (due to petroleum activities), aggregate value in agriculture was very low, lower than the state average--although sixty percent of the population worked in that sector (Gob. Est. Veracruz 1976, pp. 223-25; PEMEX 1977). Sanders' review of Poza Rica (1977) concluded that agricultural development was extremely difficult in petroleum regions.

But at both the levels of theoretical understanding and statistical evidence, most of these analyses failed to examine employment linkages between the agricultural and newly-developing nonagricultural sectors. To quote Yates (1981, p. 235), we have to visualize the Mexican agricultural work force not as a specified number of full-time workers, but as a "gigantic sponge that expands to around 10 million persons during the weeks of harvest and contracts to perhaps less than three million in the dead season."

In this context, the effect of petroleum boom-related employment opportunities on aggregate farm production depends upon the nature of the farming systems and their pattern of labor utilization. Because aggregate employment and production statistics do not measure these factors, it is extremely unclear whether those individuals whose primary occupation ceased to be agricultural did or did not continue to work in agriculture, what the disposition of their land was, and what were their prospects for future agricultural production.

Agricultural Policy During the Oil Boom

To understand agricultural policy in Mexico during the oil boom years, it is necessary to recognize that the early to mid-1970s was a period of intense national debate about the agricultural sector. To the longstanding concerns of advocates for improved campesino welfare were joined technocratic concerns over a trend of stagnation in basic food production after 1968. The 1970s was a period of experimentation with new institutions and new policies, during which agriculture was a national priority. Over-all, the share of the federal budget spent on agricultural development (a vaguely defined category) had risen steadily from an average eight to ten percent 1960-68, to 18.4 percent 1977-78, had fallen to an average 15.4 percent 1979-80 (on a much larger base), then rose again (Pirez 1983, p. 161).

For two decades, government agricultural policy had been largely directed to establishing capitalist, American-style agriculture in new irrigation districts and export enclaves (Scherr 1975; Johnston, in press; Reynolds 1970). The administration of president Luís Echeverria Álvarez (1970-76) was, by contrast, much more concerned with ejidal development. Land invasions by campesinos on holdings larger than the legal limit were more frequently condoned by the Echeverria administration. Producers (ejidatarios) on ejidos, a unique form of landholding which was the basis of agrarian reform in Mexico since the 1930s, were encouraged to adopt collectivized production systems, rather than the more typical individual family plots. The PIDER program for integrated rural development was begun and rapidly expanded to all parts of the country. A national staple food marketing system, CONASUPO, was expanded dramatically and placed in charge of agricultural trade, administration of guaranteed prices, and food marketing for poor urban consumers, including food subsidies. For the first time in many years, guaranteed prices for basic foods were raised significantly in real terms. The area on which fertilizer was applied rose from 38 percent in 1970 to 63 percent in 1978, although in 1975-80 there was no increase in the intensity of fertilizer application. There was a big increase during the decade in small-scale irrigation (CESPA 1982, Vol. X).

The Mexican government also began to make decentralization of the economy a major national priority (Pirez 1983). Map 1.2 illustrates the major regions of Mexico. Economic development had been extremely concentrated in Mexico City and the central mesa. A secondary economic center developed in the Northwest, originally motivated by the major irrigation projects of the 1950s. The highlands were the poorest areas of the country, although they remained densely popu-

Map 1.2 Mexico and the Tropical Southeast

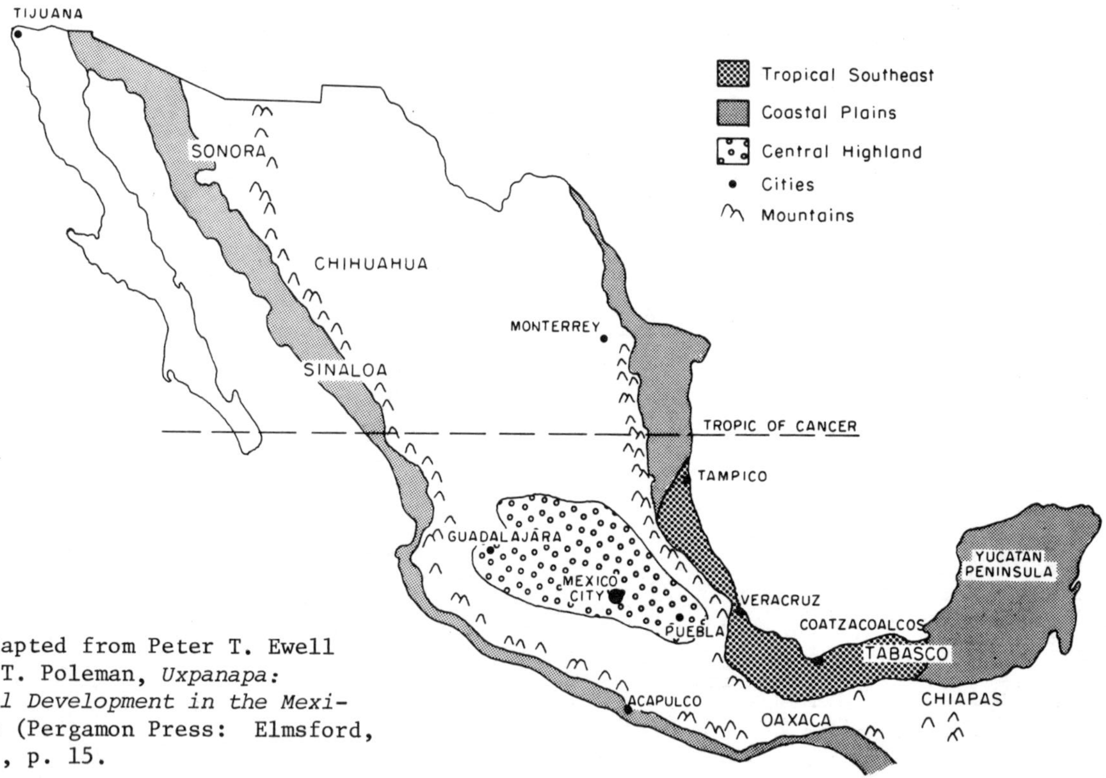

Source: Adapted from Peter T. Ewell and Thomas T. Poleman, *Uxpanapa: Agricultural Development in the Mexican Tropics* (Pergamon Press: Elmsford, N.Y., 1980), p. 15.

lated and dominated by campesino, often indigenous, populations (Reynolds 1970). The new "frontier" for economic development was the coastal lowlands, particularly in the Southeast. This region received more federal investment in industry than any other (reaching over 60 percent 1978-80); and received nearly a third of total federal investment after 1977 (Pirez 1983, pp. 158-60).

A PRODERITH report in 1978 called the coastal plains of the Gulf and Southeast zones--more than 3.6 million hectares of "high or average agricultural productivity--the most important agricultural frontier of the country" (SARH, PRODERITH, p. 3).* The region's chief attraction is its abundant water resources. During the 1970s, a major program was undertaken to transfer crops with very high water requirements out of the northern irrigation districts to substitute more valuable crops which could better justify expensive irrigation. New cane and rice processors were constructed in the Southeast; sources of oil such as coconut and African palm oil could potentially substitute for many oilseed crops grown under irrigation. As new water sources for irrigation grow scarcer and nonagricultural demand on water supplies for urban and industrial purposes becomes heavier, the need for alternative, cheaper means of production and sources of water will increase (Álvarez 1978; SARH, PRODERITH 1978).

The most important tropical products have been sugar cane, oranges, mangos, other fruit trees, cacao, cocoanut, rubber, and spices. The area in maize, rice, beans, bananas, and cassava was greater, but maize yields were relatively low, and the crop was grown largely for subsistence. Small increases in productivity could lead to greatly expanded supplies, even though basic grains can be more cheaply produced in more moderate climates. Livestock development has outpaced agricultural development in many of these regions, often on potentially good cropland, but great gains could be made in livestock production even if restricted to land unsuitable for crop production.

The administration of José López Portillo (1976-82) moved away from the rhetoric of land reform and again provided

---

*This compares to approximately two million hectares in the northern and central zones of the country which the study felt it was feasible to put under irrigation. Tabasco had one million hectares of this area, plus 600,000 hectares of average or low agricultural potential that could be used for livestock production (SARH, PRODERITH, p. 3). Yates concluded that the realistic area in the Southeast which was feasible for agricultural production was closer to 2.5 million hectares in any given year, assuming a system of soil fallow one out of four years (Yates 1981, p. 60).

strong guarantees for private commercial producers. The huge increase in food demand sparked by the oil boom was beginning to be felt by the mid-1970s, and farmers who had the wherewithal were quick to switch production to those products whose prices were rising most quickly. To keep inflation down, however, government policy kept staple food prices down as well, which accelerated the move away from maize and beans production. Remarkably rapid increases in employment opportunities resulting from the massive private and public investment and construction booms also diverted the labor of campesinos from their milpas.

Unprecedented levels of imports of maize, wheat, and soy; public concern to reduce permanent rural-urban migration to the already overcongested major cities; and the well-publicized evidence of serious malnutrition in the midst of prosperity led to the establishment of the Mexican Food Program (Sistema Alimentario Mexicano, or SAM). The SAM was announced in March 1981, two months after a major new "Law for Agricultural Development" had been passed to facilitate government programs. SAM was oriented to assist low-income staple food producers and low-income consumers through a massive system of producer input and consumer food subsidies (SAM 1979; SAM 1980a, 1980b, 1980c, 1980d; Serfin 1980; Link 1981). Almost $3 billion was spent in food subsidies in 1979, $3.7 billion in 1980. The latter figure represented nearly three percent of GDP and an estimated 12 percent of the federal budget (El Mallakh et al., p. 50). This was indeed a shifting of priorities, as the national machinery of agricultural programming and investment was redirected for the first time on a massive scale to maize producers. In addition, PIDER programs were expanded, and COPLAMAR, a program to improve welfare in marginal rainfed regions of the country, was set up on a large scale.

The record is unclear as to the exact effects of SAM--how much of the production increase was due to SAM, and who the real beneficiaries of the program were. Program funding was cut substantially after 1982, though its goals of food security, rural employment, and improved nutrition were not abandoned. As of this writing, no comprehensive analysis of the role of SAM inputs in the high production levels achieved in 1980 and 1981 has been done. Although real prices of grain continued to decline after the devaluation, relative prices improved considerably; and as competition from U.S. imports declined between 1982 and 1984 production of staple grains continued to rise. There was some evidence that agriculture policy in 1984-85 was shifting back to greater emphasis on export agriculture and capital-intensive producers (Ramírez 1983).

THE ROLE OF AGRICULTURE IN A PETROLEUM ECONOMY

Mexico's agrarian structure has been widely characterized as the epitome of agricultural "dualism" (Johnston, in press). The campesino sector, which includes ejidatarios and small property owners (propietarios) on rainfed land, produces most of the maize, beans, sugar cane, and more recently sorghum and some oilseeds. According to the 1970 Census, 87 percent of Mexican agriculturalists were "campesinos," characterized by small plot size, high employment and value of production per hectare, low value of production per person employed, and an important subsistence component (CEPAL 1982, Charts 2, 3, 34, 35, 36).

The "capitalist" sector, which includes ejidatarios as well as small private landholders (propietarios), but in much smaller proportion, has access to much higher quality land, irrigated* or "humid," with better transport infrastructure, and often a much greater extension of land. This sector also has greater access to capital and to bank credit, and produces more of the forage products, high-value fruits and vegetables, and livestock products.

A central question of agrarian politics in Mexico has been the relationship between these two groups. Modern Mexican agroindustrialists and their U.S. counterparts, as well as middle-class propietarios and ejidatarios, often take the position that the gradual capitalization of agriculture has led to rural prosperity in regions of predominantly commercial agriculture. Thus the "commercial" sector produces a disproportionate share of the value of marketed agricultural products. This group tends to believe that most land concentration, and the consequent emigration of marginal producers, is a natural and desirable part of the process of capitalist development--and that there is ample potential for development of campesino farming through new technology and institutional reforms.

Most students of the campesino economy, however, blame other factors for the difference in productivity and income between the two groups. In particular, they note the expulsion (either through market mechanisms or through force) of campesino producers from the highest quality land, and their replacement by "capitalist" farmers who monopolize available inputs and investment resources. The most critical displacement has been through accelerating substitution of pastureland for former campesino agricultural lands. The impact of this displacement has had effects not only on levels of agricul-

---

*In 1977, irrigated land accounted for 26.5 percent of harvested area, and 48.2 percent of production; these lands produced 42 percent of the nation's food for direct human consumption.

tural production, particularly basic staples traditionally produced by campesinos, but also has substantially reduced the level of agricultural employment, in conjunction with mechanization on more highly capitalized agricultural holdings. Other explanations for higher performance have included the historical bias of government investment toward capital intensive producers; cheap food policies; and rural wage levels below subsistence, which require supplementation by subsistence farming ("Alimentación" 1979; de Navarrete 1971; CEPAL 1982).

The campesino sympathizers are themselves divided about the likely course of campesino agricultural development. The "proletaristas," using a more orthodox Marxist analysis, see an unalterable process of rural proletarianization taking place and the "disappearance of the campesinado" (Barkin and Suarez 1982; "Diéz Años Después" 1983). The "recampesinistas" see the campesinado as a viable long-term force, with a special structural role on the edge of, but integrated with, the dominant capitalist economy (Esteva 1978; Warman 1980; Rendón 1975).

This study will address some of the theoretical issues of the campesinado in Mexico: the survivability of the campesino mode of production under the stress of petroleum development; the relationships between the campesino and capitalist economies; and various approaches to capitalizing campesino farms. But while the capitalist/campesino distinction is real and useful, many assumptions about their respective behaviors are not. Better field work is needed to inform the theoretical debate.

The discovery of huge petroleum reserves has in some ways changed the nature of the debate. Nonagricultural employment is now a more feasible alternative for a large proportion of the campesinos whose farms are on marginal lands. Foreign exchange is potentially available for the purchase of basic grains from the much more productive farms in the U.S. corn and wheat belts. To some observers, the agricultural problem in Mexico is today basically one of efficient and minimally disruptive transfer of a marginal farm population to the cities, and reinforcement of an increasingly capital-intensive farming system, which will be internationally competitive. despite relatively high and rising labor costs.

I believe that this view is too sanguine. Food security concerns are not irrelevant: the basically cyclical nature of a petroleum-based economy and foreign exchange availability requires that a substantial part of basic food needs be satisfied through domestic production. The fundamental development issue in Mexico--as in much of the Third World--remains the generation of remunerative employment (Scherr 1983). There is ample evidence that broad-based rural development provides more jobs per unit investment than do other sectors (Lipton

1977; Shiras 1978), particularly given the propensity in Mexico for capital-intensive industrial development (El Mallakh et al. 1984). A prosperous smallholder mixed subsistence/commercial farming sector can provide significant contributions to rural welfare, employment, urban congestion, and rural stability.

Furthermore, the extremely high rate of growth in production required to keep future imports to a reasonable level (much less reduce them) is *not* likely to come from a small capital-intensive sector only. The campesino sector must be part of any solution, and there is ample capacity for increasing campesino production, though not necessarily through the same programs which would be appropriate to the large-scale capitalist sector.

Intelligent debate about these issues is seriously constrained by the fact that there is no very clear idea in Mexico--or in most other oil-exporting countries--of the specific changes which have been wrought in the country's agricultural economy by the oil boom. We have a good idea of some of the aggregates: changes in average national prices for inputs and outputs, changes in wages and labor costs, and changes in output. The disaggregated, microeconomic information, which would provide explanations and enable us to predict future farm sector response, with few exceptions simply does not exist.

Who exactly has continued to produce? Who is increasing production? Who is decreasing production? Who is shifting to different types of products? What are the price elasticities of different groups of agricultural producers for various products and inputs? What is the effect of rural-urban migration on agricultural production in different regions? What is the effect of increasing off-farm wages on subsistence production by minifundistas, and on the process of minifundization itself?

In Tabasco, smallholders were the most productive and successful farmers, even in production of high-value crops, and those most capable of adjusting to the economic pressures of a petroleum boom. This evidence suggests the need for a re-evaluation of the role of mixed subsistence/commercial producers in economic development, particularly in a petroleum economy. Campesinos may play a potentially more dynamic role in stabilizing and developing rural economies under stress than is commonly believed.

The next chapter reviews the history of Tabasco state development, highlighting the interactions between the agricultural economy and the petroleum industry during the decade of the 1970s. This history confirms the depth of change and disruption which petroleum brought to Tabasco, and also documents the state's unexpected agricultural boom and the principal factors responsible for that boom. Then, using the cacao sector as a case study, Chapters 3-8 will describe and analyze the interactions among farmers, farmworkers, PEMEX, government programs, and the general Tabascan economy during the period of the oil boom.

2

An Introduction to Tabasco: History and the "Boom" Experience

In the Chontalpa region, there is too much oil. Unavoidably a very serious conflict between agricultural development and the petroleum industry will take root. Who will vanquish whom? Undoubtedly it is petroleum that is winning the battle up to now.

--Manlio Tirado, 1977
[my translation]

Basically, major conflict between petroleum and agriculture is unnecessary, if enough support is given the agricultural sector. We are doing that. Look at the results: the highest cacao and coconut production in history.

--Governor Leandro Rovirosa Wade
Tabasco, 1979 [my translation]

TABASCO: AN ARCHIPELAGO

Tabasco lies on the alluvial plain below the Isthmus of Tehuantepec, stretching along the Atlantic side of the Sierra Madre. The fundamental feature of its geography is an abundance of water. Only slightly exaggerating, one observer noted: "It is the only case in the world of an archipelago situated on solid ground" (Bolís Villanueva 1976, p. 3). The Tabasco plain receives the flow of Mexico's mightiest rivers --the Usumacinta and Grijalva--which together with their also impressive tributaries, pour into this state about one third of all the water resources of the country.

Map 2.1 shows the principal features of the state and its major sociogeographic regions: the Chontalpa, Rivers, Sierra and Center. Maps 2.2 and 2.3 contrast the natural vegetation of Tabasco with contemporary land use patterns.

The mountains south of Tabasco form a barrier to moisture-laden Atlantic winds, resulting in a rainfall pattern which ranges from an average of only 1,500 mm per year along the coast to nearly 5,000 mm in the peaks above Teapa: the second rainiest spot in the world, according to local claims. The rainy season lasts from seven to nine months, lengthening nearer the Sierra. For at least eight months of the year, 52 percent of Tabasco's land area (mostly in the Rivers region) is covered with water; thus the description "archipelago." Fifteen percent is permanently in swamp.

The rivers carry a remarkably huge burden of sediment each year, much of which is left along the floodplain, enriching the already deep alluvium, high in potentially fertile clay soils. On the natural dikes of the Old and New Chontalpa, this forms exceptionally good soils; where the ground water is higher and in the depressions (such as the Chontalpa "Bowl"), soils are often gleyed and have formed impermeable horizons.

The Rivers region, whose natural vegetation was either swamps or mixed forest and savannah, is now used for extensive cattle pasture in the dry season, or year-long, depending on drainage. The Sierra region, originally mixed savannah-forest, with rainforest in the southern mountains, is also primarily a cattle zone. Along the Teapa and Tacotalpa river banks, however, are fertile, well-drained soils which can be used for crop production. Like the southern Center region, it is particularly suitable for banana production. Because of higher rainfall and cooler temperatures, the area is less suitable than the Chontalpa for cacao production.

The Chontalpa is the agricultural heartland--and hence historic population center of the state. Except for the mangrove forests and swamps along the coast, now used for coconut plantations, the natural vegetation of this region was rain forest. But the rain forest has disappeared in all but the most remote corners of the state. On well-drained land, the cacao plantations with their shade trees mimic the natural forest structure, protecting soils from severe erosion and building fertility. A series of huge drains built in the 1950s and 1960s as part of the Chontalpa Plan opened a large area formerly subject to flooding, to sugar cane, banana, and year-long pasture production.

On other lands, deforestation occurred over many centuries due primarily to burning for agricultural uses. Exposure to leaching and high temperatures led to decreased fertility over large areas and the less fertile, mixed-savannah-forest

Map 2.1 Principal Towns and Road Communications in Tabasco State, 1980

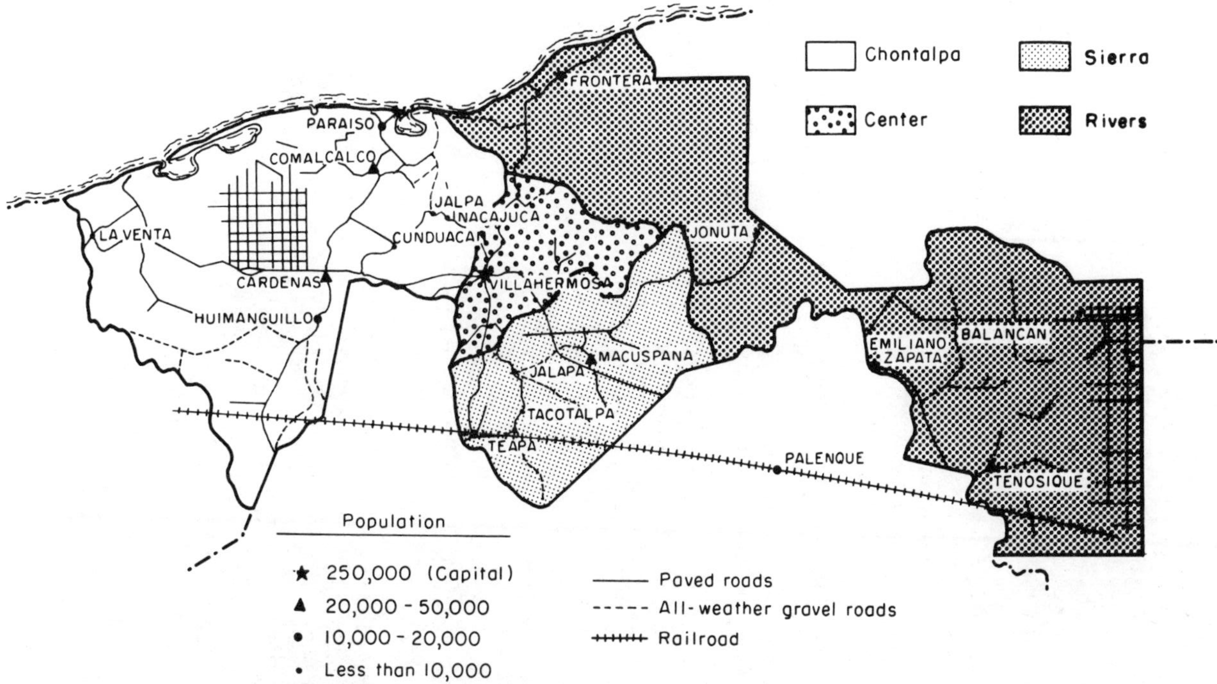

Source: The road system is a revised version of that found in Map 4 of Eduardo Bolís Villanueva, *La Economía del Estado de Tabasco*, Banco de Comercio, March 1976. Town population figures were taken from preliminary data of the 1980 National Population Census.

Map 2.2 The Natural Vegetation of Tabasco

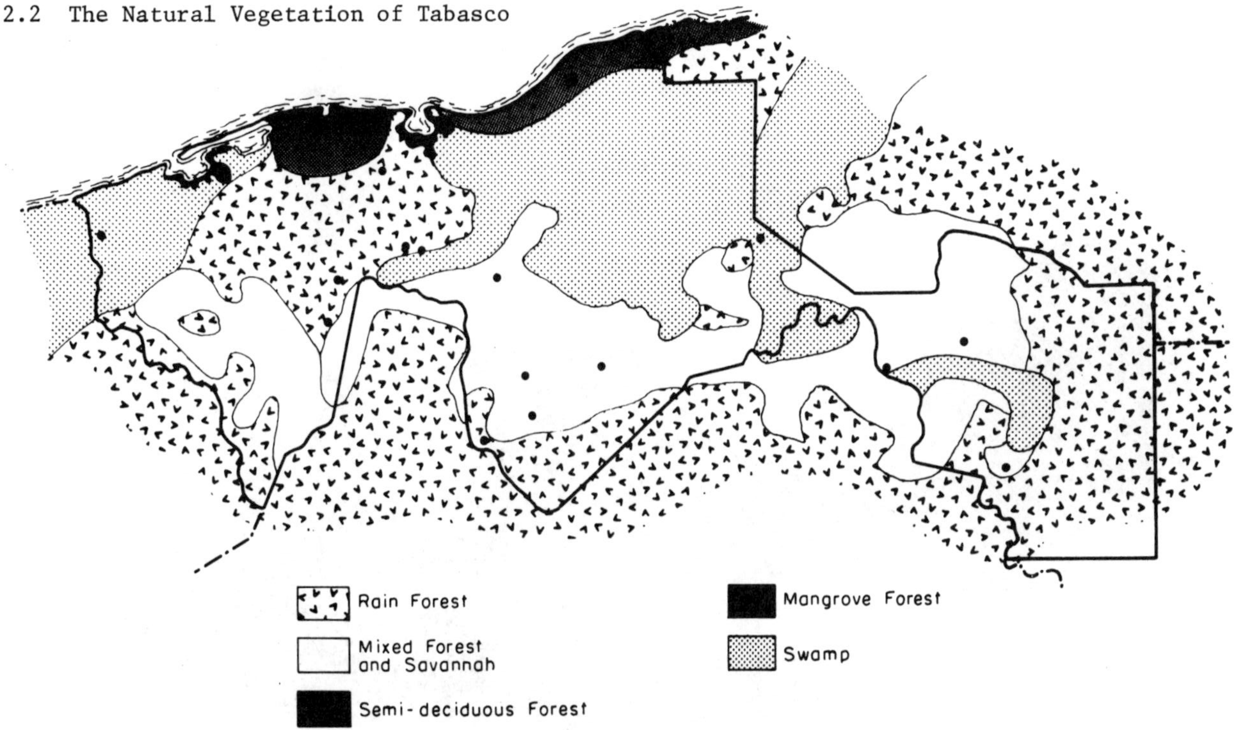

Source: R.C. West, N.P. Psuty, B.G. Thom, *Las Tierras Bajas de Tabasco en el Sureste de Mexico*, Gobierno del Estado de Tabasco, Villahermosa, Tabasco, 1976, p. 76.

Map 2.3 Contemporary Land Use in Tabasco

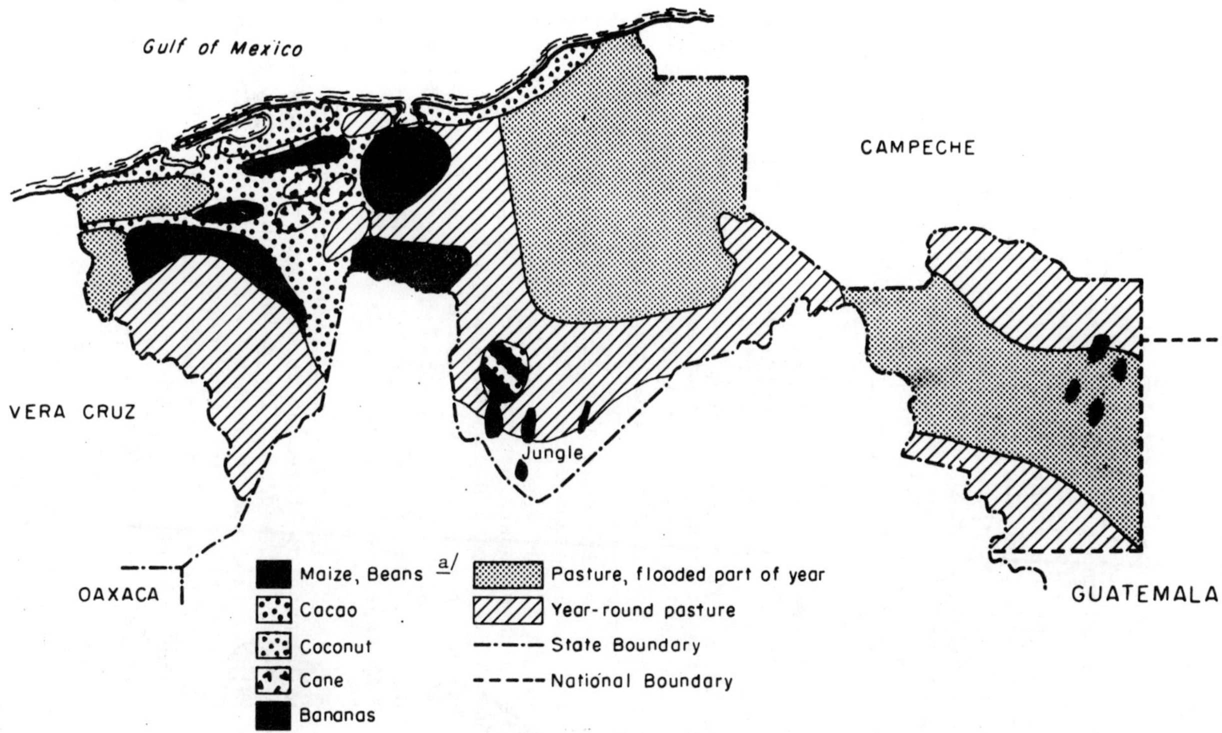

[a]Note that additional subsistence maize and beans are found throughout the populated regions of the state.

Source: Based on SARH, Representacion de Tabasco, "Plan Agrícola Estatal," 1979, Land Use Maps.

land turned to pure savannah. These soils, primarily found in the Savannah subregion, are now used for cattle production.

Traditional Agriculture in Tabasco

Archeological ruins indicating early Mayan and Olmecan settlements can be found throughout the Old and New Chontalpa and all along the Usumacinta River. The indigenous population at the time of the Conquest was predominantly Chontal, with some Nahuatl, Popoluca, and Zoque populations. They were concentrated in the Old Chontalpa and less so in the Sierra, along well-drained natural river dikes where the best cacao land was found. Indigenous population estimates for the time of the Conquest run as high as 300,000, but the most likely figure was 135,000 (West, Psuty, and Thom 1976, pp. 114-15).

Subsistence was obtained through yearly double-cropping of maize, often with slash and burn, or in flooded areas, through seasonal movement from flooded to unflooded soils. The main staple was pozol, a thick liquid mixture of crushed maize and cacao; accompanying foods were the beans, squash, tomatoes, and chiles typical of Mexico, plus a number of tropical tubers not commonly consumed farther north, such as sweet potato and cassava.

Cacao was grown in Tabasco as early as the fourteenth century, and the Tabascan cacao variety "criollo" is considered the origin of the species. At the time of the Conquest there were thick stands along what are today the Cunduacán River and the dry riverbed of what was the San Bernabé River. Seventy percent of the cacao was located in the Chontalpa; the rest in stands along the Grijalva and Usumacinta rivers, near Villahermosa, and to the south along the Teapa, Tacotalpa, and Chilapa rivers (West, Psuty, and Thom 1976, p. 130).

Cacao was used as the medium of exchange for trade between the Mayan city states of Yucatan and the Caribbean coast, and the Aztec kingdom. The principal trade cities --Xicalongo, Cimatán, and Potonchán--were usually controlled by Aztec or Mayan traders, who handled salt, slaves, cotton cloth, tools, and honey from the Guatemalan coasts; decorated cloth, gold and turquoise objects, copper bells, dyes, obsidian tools, and slaves from Central Mexico (West et al. 1976, p. 116).

Tabasco also sent cacao--a luxury good from which a foamy chocolate beverage was made--as tribute. One document from 1562 reports that 81 tons of dry cacao was sent from Tabasco annually to Veracruz and Mexico City--30 percent as tribute, the rest for commerce (West et al. 1976, p. 128).

DIVERSIFICATION OF THE TABASCAN AGRICULTURAL ECONOMY

Between the indigenous period and the modern era, the landscape and economy of Tabasco changed radically.

The Colonial Period: Cacao, Cattle and Dyewood

After the Conquest, Bernal Díaz himself sought an encomienda of cacao plantations in Tabasco. But with the coming of the Spaniards, the coastal population was decimated through diseases and violence. At the end of the eighteenth century, the total population of Tabasco was probably no more than 10,000 (Gerhardt 1979, p. 25). The Spaniards themselves fled from the lowlands, which were unhealthy and prone to attack by pirates, to the Sierra where they and mestizos replaced most of the Indian population. The state capital moved from Santa María de la Victoria to Villahermosa in 1602, then to Tacotalpa in 1666, and back to Villahermosa--today's capital --in 1796. To fulfill labor needs on the cacao plantations, salaried Indian workers from Chiapas were brought to Tabasco; some stayed, others moved back and forth seasonally.

Tabasco continued to produce cacao. But import competition was fierce from Ecuador and Venezuela, and later from other plantations established by the Spanish in the Caribbean and Central and South America.

Seeking other sources of income, the Spaniards introduced two far-reaching economic changes. Cattle were brought in and served to open interest and settlement in the interior savannahs of Tabasco and the fresh water swamps and fields by the coast. Cattle, skins, and tallow became important sources of income for local Spaniards.

The second major change was the exploitation of forest resources, mostly the "pita," a tropical fiber; sarsparilla, a medicinal root; Jamaican pepper; and dyewood, a tree which produced a deep red to purple dye, which was by far the most important. Spaniards participated in this trade, but after 1660, English "semi-woodsmen, semi-pirates" began active exploration for dyewood, particularly around the Laguna de Términos.

Independence: Migration to New Lands

After Mexican Independence in 1821, the main process leading to change in Tabasco came from migration to previously sparsely settled land. This took place in three movements: from Old to New Chontalpa, from Cárdenas northwest, and from Huimanguillo east. The population of Tabasco at that time is

Map 2.4. Historical River Pattern Changes in Tabasco.

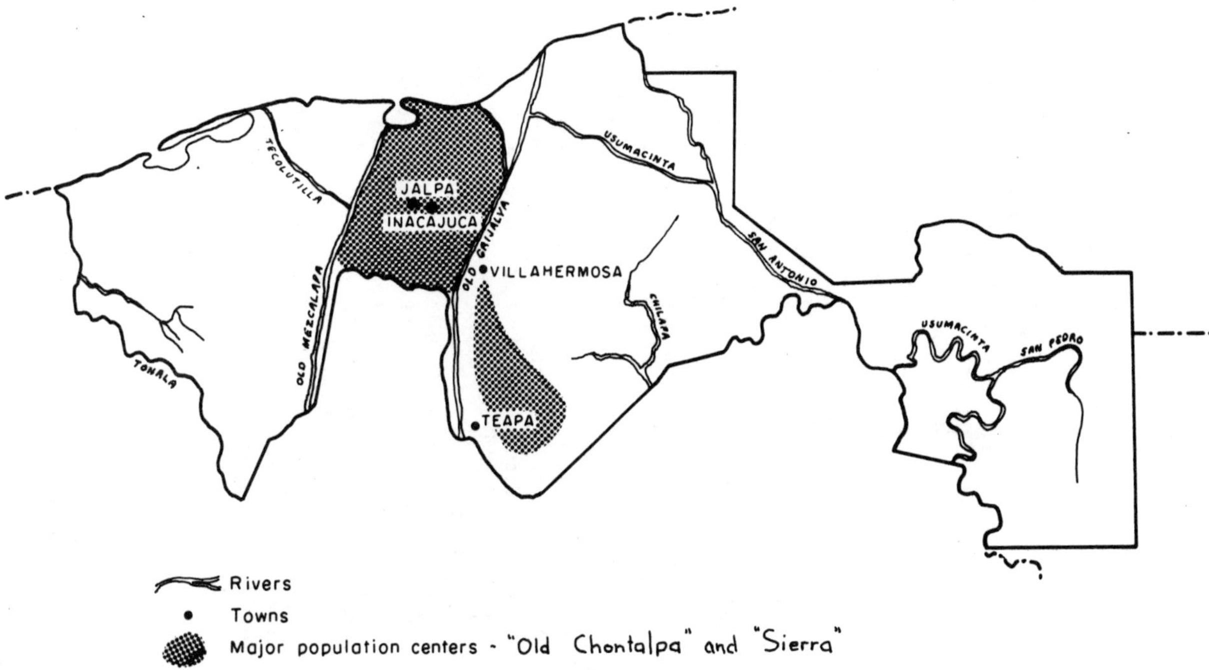

A. Pre-1675 River Flow Patterns

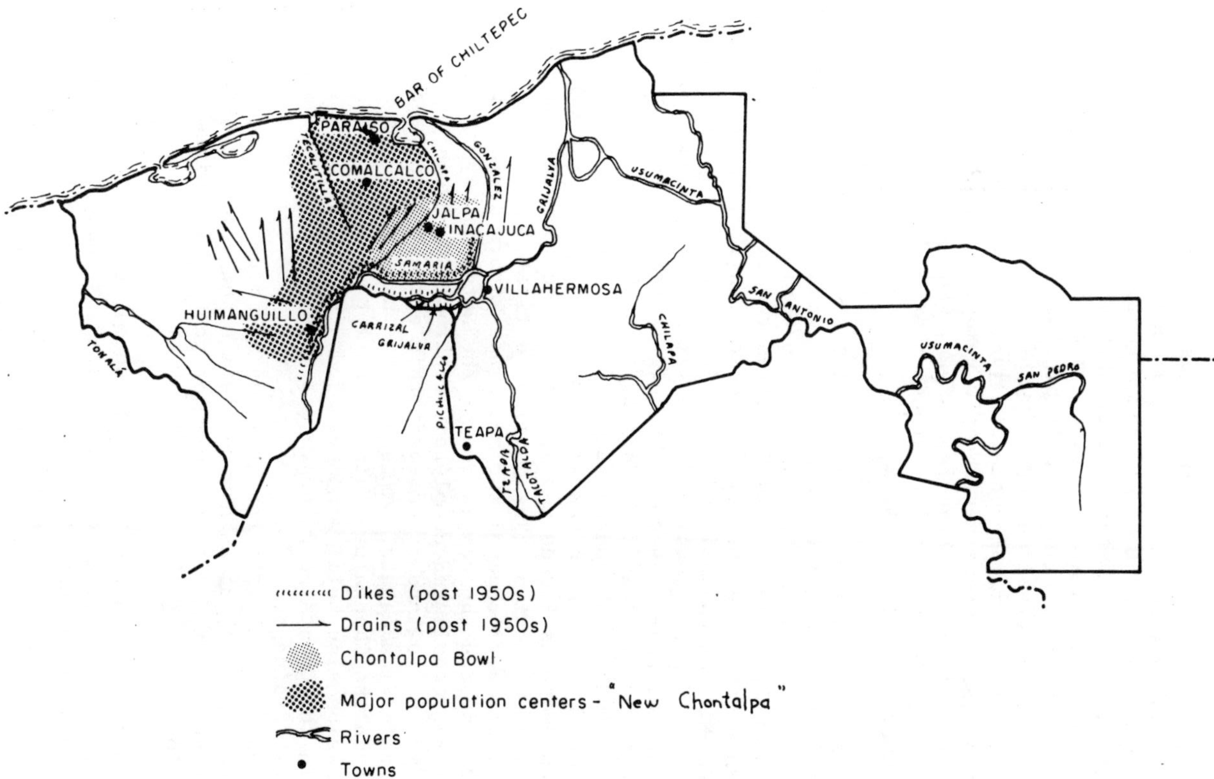

B. Post-1932 River Flow Patterns

[a]The rivers that flow into Tabasco from the Sierra are both voluminous and precipitous, a result of moisture-laden rains from the Atlantic falling on geologically new mountain formations. These waters are quickly slowed as they reach the near-level plain of Tabasco, which offers few natural barriers to water flow. The combination has meant relative instability of river channels in the state. The piecing together of historical documents suggests that since the time of colonization there have been a number of major changes.

Before 1675, the Mezcalapa River, whose origins are in Guatemala, flowed directly through the Dos Bocas River (whose now empty channel is called Río Seco) to the sea. It was a completely separate system from the Grijalva-Usumacinta. In that year a "rompido" (break) in the channel diverted the river east, connecting the Mazcalapa with the Grijalva system, near what is now the Chiapas-Tabasco border.

In 1881, a new "rompido" created the Carrizal River, which flowed north to the Bar of Chiltepec. In 1902 the Carrizal abandoned that channel and again joined the Grijalva. In 1932, the Samaria rompido formed yet a third channel connecting the two river systems. The new Samaria River emptied through the old González, Carrizal, and Grijalva rivers.

The 1932 break caused severe flooding in the Chontalpa Bowl area. Another break in 1952 in the Carrizal River also caused enormous floods and stimulated the formation of the Grijalva Commission. The commission first constructed artificial dikes to channel the Mezcalapa waters during flood. After 1966 and the completion of the Netzahualcoyotl Dam, the river was considered completely controlled and has not since flooded the lowlands. However, this did not resolve the problem of inadequate drainage of the huge volume of rain that falls on the Chontalpa plain in the wet season. Thus, the Commission also built a number of drains in the Chontalpa Bowl and in the Chontalpa Plan project to supplement the poor natural drainage.

Map 2.4 contrasts the pre-1675 and modern river channel patterns and indicates how population and economic centers of the state shifted with these major geographic shifts. See text for more detail.

Sources: Eduardo Bolís Villanueva, *La Economía del Estado de Tabasco*, Colección de Estudios Economicos Regionales, Sistema Bancos de Comercio, S.A., Mexico D.F., March 1976, pp. 8-9; R.C. West, N.P. Psuty, and B.G. Thom, *Las Tierras Bajas de Tabasco en el Sureste de Mexico*, Gobierno del Estado de Tabasco, Villahermosa, Tabasco, December 1976, pp. 183-85.

estimated to have been nearly 60,000, of whom half were Indians (Gerhardt 1979, p. 25).

The Chontalpa movements were occasioned principally by major changes in river flows, which are diagrammed on Map 2.4. Originally, the Grijalva and Usumacinta systems were entirely separate, with what is now called the Grijalva actually the Sierra River. What is now called the Mezcalapa was the Old Grijalva River.

Until 1675, the Old Grijalva flowed nearly north to the sea. After that date, it abruptly turned east, joining the two systems. In 1882, its course broke again to form the Carrizal River, which flowed to the Bar of Chiltepec (through the old González River) rather than the Grijalva-Usumacinta system. In 1902, it was diverted once again to the new Grijalva; and in 1932 opened a third west-east branch--the Samaria River--whose water flowed north at times to flood the Chontalpa Bowl. The Samaria River reached the sea through the Bar of Chiltepec, the Carrizal and González rivers, and the new Grijalva-Usumacinta system.

The major effect of these changes in river course was to reduce the productivity of the Old Chontalpa region, while opening up the extremely fertile, now dry riverbed of the Old Grijalva River--the "Río Seco." Massive emigration began from the former to the latter for cacao planting. In 1823 whites and mestizos settled the town of Paráiso and in 1827, Comalcalco--right in the old riverbed. By 1854, the region was prosperous and full of cacao (West, et al. 1976, p. 141).

The second migration stemmed from the town of Cárdenas, which had been established in the old Río Seco at the end of the eighteenth century. By 1869, 1,159 Indians and mestizos had moved along the old tributaries to the Mezcalapa to plant cacao and subsistence crops.

The third route of colonization was of Ahualculco Indians who settled near Huimanguillo and Ocuapan, at the close of the eighteenth century along major rivers west, toward the present border with Veracruz, which their presence established in the 1800s.

These changes brought renewed vigor to cacao activity in Tabasco. Whereas in 1854, the Sierra had three quarters of the cacao trees in Tabasco, by 1890, cacao production in the Chontalpa had once again surpassed that of the Sierra, which had suffered severely from floods and droughts in the interim (West, et al. 1976, p. 162). In 1828, Mexico's first chocolate factory using sugar was built; in 1876, the first milk chocolate factory (Penny 1973, p. 10).

After the Great Dry Spell 1859-61 caused a sharp decline in cacao production in the Sierra, large sugar cane haciendas were planted for the first time on the floodplain of the Tacotalpa, producing rum and raw sugar. Hevea rubber, coffee originally used as cacao windbreaks, and some tobacco near

Ocuapan were also planted in the late nineteenth century in small quantities. The expansion of cattle production, and a vast increase in logging stimulated by United States demand for mahogany, exterminated most of the remaining jungles, completing the land use changes which took place between Independence and the Mexican Revolution.

Early Twentieth Century: Coconut, Bananas, and Improved Cattle Breeds

Tabascans were active in the Revolution, largely because of the considerable local resentments against largeholders--primarily owners of cane and cattle ranches--who practiced debt peonage and encroached on smallholder land. Nonetheless, new ejidos promised by the federal government did not take form until the 1950s.

The major changes in the first half of the century were the introduction of bananas and coconut and the spread of cattle ranching. The most famous of the post-Revolution governors of Tabasco, Tomás Garrido Canabál (1928-34), a radical populist, organized farmer producer groups for all major products, which have remained active to the present.

At the turn of the century, several landowners, inspired by the lucrative production of bananas in Central America, began commercial production of the "roatán" or Gros Michel variety on the well-drained river dikes. At first they cleared previously forested land, but later they replaced food, rubber, and cacao crops and encroached upon remaining smallholder properties all along the Usumacinta. By the 1920s and 1930s, a "banana boom" had spread throughout the lower Grijalva, the Sierra rivers, and even the Chontalpa. Marketing and transport were organized by the newly formed cooperatives and exports were shipped to the United States by boats belonging to the large U.S. fruit companies.

Between 1920 and 1930, bananas replaced cacao as the primary source of agricultural wealth in the state. During 1938-42, however, sudden attacks of the Sigatoka disease (a banana leaf disease) and Panama disease (a root parasite), coupled with lack of space in shipping boats during World War II, led to a drastic decrease in production. Through copper sulfate applications for Sigatoka control, the industry recovered in the 1940s but never resumed exports. Usumacinta and Sierra plantations reverted to livestock and subsistence crops; only the Center and Southern Chontalpa zones remained productive (West, et al. 1976, p. 168).

The first coconut palm was planted on the Chiltepec and Frontera coasts at the end of the past century, but major production did not begin until World War II, when the normal copra supplies from Asia and Oceania were cut off from the

Allies. High prices led to large plantings all along the coast--the first intensive use of this land, which had formerly been utilized only for pasture--and the first population movement there (West et al., p. 170).

Until the 1930s cattle herds in Tabasco were almost entirely "criollo." In 1932, Brahman cattle were imported, and later Santa Gertrudis, part of a new breeding program of the Cattle Producers' Association. Improved pastures had been brought in earlier: Egypt grass (*Panicum purpurascense*), 1872; Guinea (*Panicum maximum*), 1900; molasses grass (*Melinis-minutiflora*), 1923. To permit higher stocking rates of the new crossbred cattle, other new pastures were introduced; elephant grass (*Penniseum purpurerum*), pangola (*Digitaria decumbens*), jaragua (*Hypererehenia rufa*), and German grass (*Echinochloa polystachya*).

Land Use Changes after 1950

Middle-aged Tabascans still remember when one prepared for a trip to Mexico City by writing one's will. From the town of Teapa, for example, two to three days in a small boat brought one to the sea, where a precarious sea voyage to Veracruz was followed by a dizzying climb overland past the peaks of Orizaba before descending into the Valley of Mexico.

But in 1950, the National Railroad of the Southeast was extended through Teapa and Tenosique on the higher, more stable land above the floodplain, to avoid flood risks. Since population centers were located far north of the tracks, the rail line stimulated building of numerous north-south roads. While the railroad was being built in 1948, the state's first all-weather road was built from Villahermosa to Teapa, and a road between Puerto Ceiba and Huimanguillo, through Comalcalco, was gravelled. In 1961, a paved road connected Villahermosa with Mexico City. In 1967 Villahermosa was finally connected by an all-weather paved road to the Yucatan peninsula.

Flood Control. With increasing commercial activity on the Tabascan floodplain, the age-old problem of flooding became an important political issue. In 1932 and 1952, the Old Chontalpa suffered devastating floods that, for the most part, turned what had been the second most important producing zone of the state into an area of subsistence agriculture. In response to the disasters, the national government set up the Grijalva-Usumacinta River Basin Commission in 1951 to control flooding in Tabasco and to promote general economic development there.

The Commission completed the Netzhualcoyotl dam at Malpaso, Chiapas in 1964, and dams at Chicoasén and Angostura in the 1970s, whose primary role was electricity generation.

Large drainage structures built to control the Mezcalapa waters shown on Map 2.4 (p. 37). The Old Comalcalco road, the Left Dike and Right Dike protect the Río Seco area and the Villahermosa-Cárdenas highway, at the expense of increased flooding in the western zone and the "Bowl." The Samaria, Veladero, and K-24 Drains protect much of the area between Cunduacán City and the highway, but they have been continually threatened with siltation. The projects have protected the New Chontalpa area and the city of Villahermosa from further serious flooding.

A series of drains were built west of Cárdenas in the late 1960s for the "Plan Chontalpa" to foster intensive agriculture in that formerly ill-drained zone. Expensive infrastructure projects including paved roads, villages with homes of concrete block and full water, sewerage, and electricity services were carried out for 22 newly-formed ejidos.

Truly effective river flood control--even if possible, given rapid silting due to the rivers' heavy sediment loads and potential shifts in river flows--would depend on the construction of several more dams. Meanwhile, the major source of flooding in the state is the rising water table resulting from extremely high rainfall and poor natural drainage. This characteristic is a major factor in livestock management in the state. Considerable land which would be suitable for crops is used for grazing during the height of the rainy season; otherwise the cattle, which normally graze in the lower regions unsuitable for crops, would have to be sold yearly (Yates 1981, p. 58).

Growth of Commercial Agriculture. The improvements in transport, land reform, and drainage opened new opportunities for commercial agriculture in Tabasco for both formerly subsistence-oriented campesinos and market-oriented farmers. Total land in production rose greatly in the 1950s and 1960s, with livestock production becoming more and more important. Between 1965 and 1972 the value of total foreign exports from Tabasco ranged from $45 to $157.8 million--ten to 20 percent of the total value of agricultural production, largely in cacao, coconut, and shrimp. The state's overall trade balance was positive and high. However, crop prices during the 1960s were generally stagnant, and there were no correspondingly impressive gains in productivity.

Just between 1965 and 1974 the area in cropland rose by 30 percent, although crop agriculture in 1970 still accounted for only seven percent of total land use in Tabasco. Thirty-six percent was in cultivated pastures; 13.3 percent in natural pastures; 11.1 percent in forest land, 4.8 percent uncultivated productive land; and 17.6 percent was unclassified, most of which was probably flooded lands. In 1974 the nine leading crops, and their respective areas, in hectares included: maize (67,005); cacao (40,000); coconut (25,000); sugar cane (13,343); rice (12,925); bananas (7,962); beans (5,822); coffee (1,741); and oranges (1,235) (Bolís 1976, Charts 10, 11, 12).

Ejidos controlled over a third of all farmland, more in the agricultural zones of the Chontalpa. Private land ownership in Tabasco was extremely concentrated, with four percent of owners controlling 59 percent of private agricultural land. Half the private farmers (with only three percent of the private farmland) held under five hectares (SIC 1970). However, most of the large extensions were in the Rivers area, the Sierra, or the Savannah; in the rest of the Chontalpa, there were few large properties. In fact, the most important agrarian problem there was minifundization, which had been accelerating since the final settlement of the frontiers earlier in this century.

A close interface between cattle farming and the smallholder sector developed in the 1970s. The growth of the cattle industry led to greater demand for forage, so that ranchers began to contract with ejidatarios and other smallholders for grazing rights in return for keeping half of the year's new calves. This is called "medianía" and is distinct from the other common practice of simply renting pastureland for a fixed price per hectare per month. Promoted by government loan programs and medianía, livestock are now an important asset of cacaoteros and other smallholders.

PEMEX IN TABASCO

PEMEX in Tabasco before the "Boom"

Modest drilling in Macuspana and Reforma, Chiapas by private oil companies and PEMEX before 1950 had little impact on the state (West et al. 1976, p. 178; "Producción" 1977). But in 1951, the most productive site in the state known before the 1970s was discovered in Macuspana, where PEMEX City, an industry "new town" for 300 employees was set up. Pipelines to send gas to Minatitlán, Veracruz, and Mexico City were built. Comalcalco was made District headquarters. Between 1954 and 1958, new sites were found near La Venta, related to the formations of southeastern Veracruz. A second permanent establishment was set up in La Venta by PEMEX, and dozens of temporary "lost cities" sprung up.

The Isthmus salt basin near the Veracruz border also produced some oil and gas; in the area stretching from Macuspana northeast to Campeche, huge gas reserves were found. The first commercial production from Gulf off-shore areas was in 1958. By 1961, the entire Southern Zone, including the

Isthmus with 42 percent of the nation's wells, was producing 24 percent of Mexican crude, 24 percent of natural gas liquids and 44 percent of the natural gas (PEMEX 1978b; 1979).

The Oil Boom of the 1970s

In 1970 PEMEX accelerated its drilling activities in Tabasco and began to work much deeper, at 4,500 meters. With wells Sitio Grande #1 and Cactus #1 in 1972, and Pozo #5 in February 1973, they struck high pressure pay ("Producción" 1977, pp. 4-5). Subsequent exploration wells were drilled with extraordinary success rates and yields per well. The oil boom had begun.

After 1976, PEMEX expanded rapidly. Between 1976 and mid-1978, the number of productive wells rose from 114 in seven fields to 160 in 14 fields, with expectations that by 1983 there would be 477 wells drilled (Tirado 1978, pp. 10-11). These areas are shown on Map 2.5 along with other petroleum infrastructure in the zone, and newly planned projects.

A petrochemical plant was built in Cactus in 1973-74 for gas-sweetening and expanded to a capacity of 200 m. cubic feet per day by 1979. Gasoline pipelines to Cactus from Cunduacán, Sitio Grande, and Níspero and 368.8 kilometers of polyducts were built (including the start of the Reynosa-Cactus gas pipeline). A vast network of roads was built through the swampland, 225 kilometers by 1978. Future plans included expansion of the Cactus plant and construction of a new petrochemical center in Gil Perez; a port facility in Dos Bocas capable of exporting 700,000 barrel/day oil; and a 6,500 ton/day capacity ammonia plant in Villahermosa, in addition to more pipelines and wells (Sanders 1977, p. 6).

PRINCIPAL EFFECTS OF PEMEX ON THE TABASCO ECONOMY

PEMEX invested billions of pesos in Tabasco during the 1970s, building wells and other infrastructure as rapidly as possible. This activity disrupted Tabascan society in four principal ways: inflating the economy and causing commodity and labor scarcities; stimulating migration and changes in the labor market; subjecting agricultural land to expropriation, trespass and pollution; and disrupting established social patterns and mores. The impact was felt most strongly in the quadrangle formed by Villahermosa, Cunduacán, Cárdenas, and Reforma towns. The effects in the Rivers, eastern Sierra, and Savannah regions were minimal.

Map 2.5 Petroleum Infrastructure in Tabasco, 1980

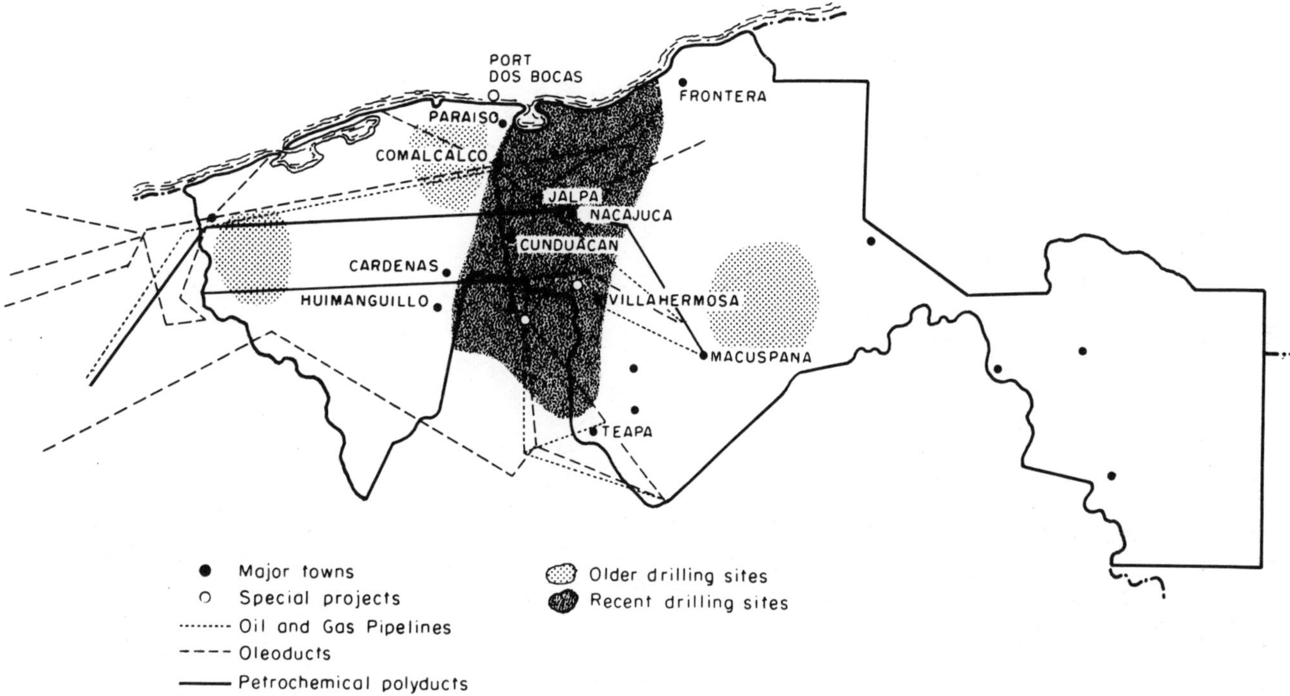

Source: Mexico, Secretaría de Programación y Presupuesto, *La Industria Petrolera en Mexico*, 1979, pp. 353-55.

Economic Impact of PEMEX

Total national petroleum investment rose from $4,576.1 million in 1971 to $62,703 million in 1978. Of this, the greatest part was spent in the Tabasco/Chiapas area. Chart 2.1 shows changing federal investment in Tabasco between 1959 and 1979 and the dominant role of PEMEX. In 1978, PEMEX paid $7.2 million per day in salaries in Tabasco. By contrast, the total value of agricultural and livestock production in the state in 1974 was only between $1,500 million and $2,000 million (COPRODET 1979a).

This profile is somewhat misleading, as most PEMEX expenditures in Tabasco were for purchases made elsewhere. An estimated 15.5 percent were for local wage and salary payments, and 21.9 percent were for profits and other expenses in the state (COPRODET 1979a, p. 59). Most internal economic linkages were for construction work.

PEMEX became the most important taxpayer in Chiapas and Tabasco. Total state and municipal receipts in Tabasco rose markedly (COPRODET 1979e, p. 14):

1973	$ 1.5 m.
1974	14.0 m.
1975	150.0 m.
1976	280.0 m.
1977	650.0 m.

By comparison, in 1976 the state received only $21.3 million from the cacao sector and $81.5 million from the livestock sector (Tirado 1978, p. 63). Even after the oil boom subsided temporarily in 1982, Tabasco's revenues, based on levels of extracted petroleum, remained very high.

This enormous increase in local purchasing power stimulated the commercial and service sectors. The number of established commercial enterprises rose 30 percent between 1970 and 1978 (compared with 65 percent 1960 to 1970 for smaller-scale operations). The volume of commerce rose many times this, even without including nonestablished commerce (such as street vending), which was evidently booming.

Inflation and Scarcities

As a result of this increased economic activity introduced from outside the state, inflation had run rampant. Some items, particularly construction goods, ran 70-100 percent higher in cost in Villahermosa, Cárdenas, and Comalcalco than in Mexico City in 1979-80 (COPRODET, 1980c; Gob. Est. Tabasco 1979a). Based on their typical "food basket," the cost of food for a family in Cárdenas town rose between 22 and 51 percent between 1976 and 1979 (Hernández 1978).

Chart 2.1 Federal Investment in Tabasco, 1959-79 (by presidential term)

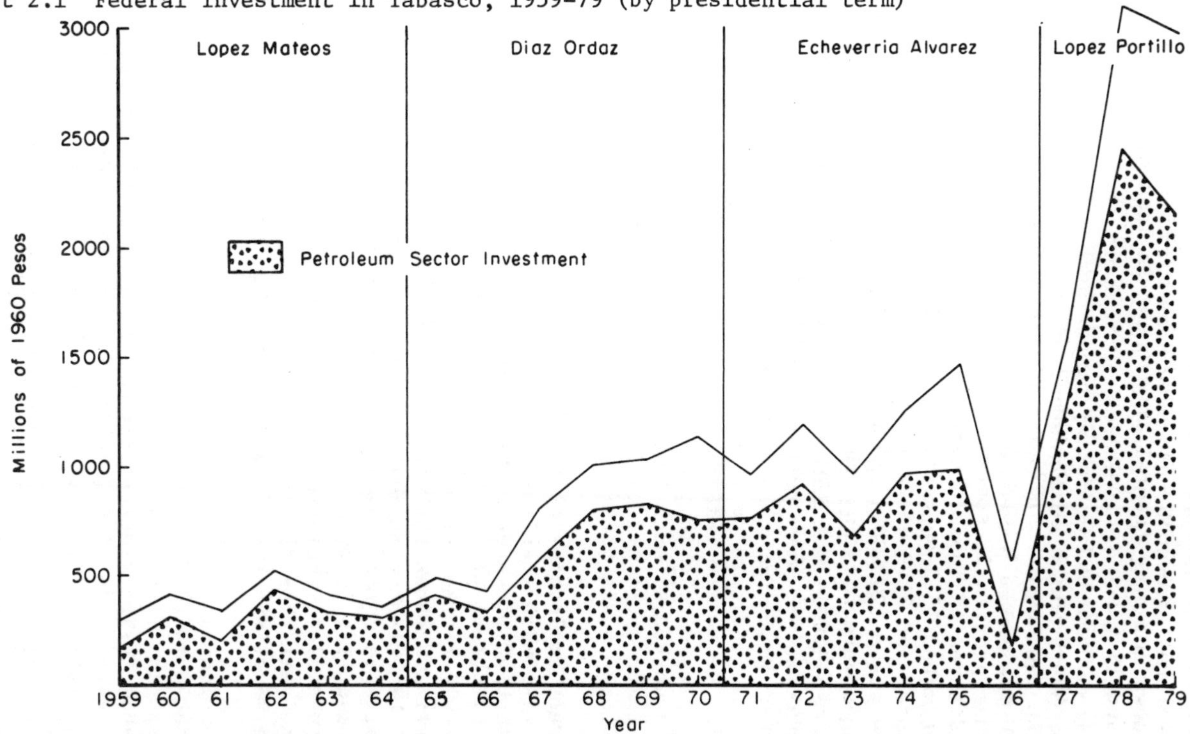

Source: Tabasco, Comité Promotor del Desarrollo Socioeconómico del Estado de Tabasco, *Análisis de las Inversiones Públicas 1959-1979*, Charts 6 and 9.

Housing was particularly scarce in the major towns and extremely expensive, even in squatter settlements. Most available hotel rooms were rented by working people, disrupting the state's tourism program for lack of accommodations.

The state government had made some strides in controlling specific types of inflation, such as food (by building CONASUPO and cooperative stores), transport (building new trucking centers and operating competing bus lines), and housing (operating a small number of direct and contracted building programs). But the devaluations of 1982 exacerbated inflationary pressures.

Labor Market Disruptions

In general, wages in Tabasco have increased faster than the cost of living index in the past decade (Scherr 1983b, Table 1). PEMEX basic labor payments were well ahead of most other groups. To work full-time ("de planta"), one had to be a union member and pay an expensive entrance fee. The union--the Sindicato de Trabajadores Petroleros de la República de México (STPRM)--was extremely powerful and played a large role in running PEMEX. Temporaries ("transitorios") worked at much lower rates, yet they were chosen by the union whose officials often charged for bestowing jobs. Large-scale contractors also frequently hired through the union.

Direct employment by PEMEX and the construction companies contracting with them rose from about 6,000 in August 1977 to between 40,000 and 50,000 in 1979-80 (Tirado 1978, pp. 13-14). They lived primarily in Villahermosa and Cárdenas and scattered near the petroleum operations.

Chart 2.2 shows the composition of the EAP of Tabasco in 1970. The entire population working in construction at that time was only 6,000. Agriculture was obviously dominant. Very few workers earned over M$1,000 monthly--a few workers in large-scale transformation activities, oil workers, a few thousand participating in high-income service activities in Villahermosa, and a very small percentage of farmers.

By 1980, the proportion in oil-related activities had grown enormously, as had the construction sector as a whole and urban services. Agricultural employment appeared to be much less important. The percentage of the EAP 12 years and older whose primary occupation was in agriculture declined from 62 percent in 1970 (SIC 1970) to 43 percent in 1980 (Negrete 1984, p. 95). In absolute numbers this represents a decline of about ten percent in the agricultural EAP.

These statistics, however, indicate only the primary occupation. The Sociodemographic Survey of Tabasco showed that secondary employment was widespread. It was also

Chart 2.2 Characteristics of the 1970 EAP in Tabasco[a]

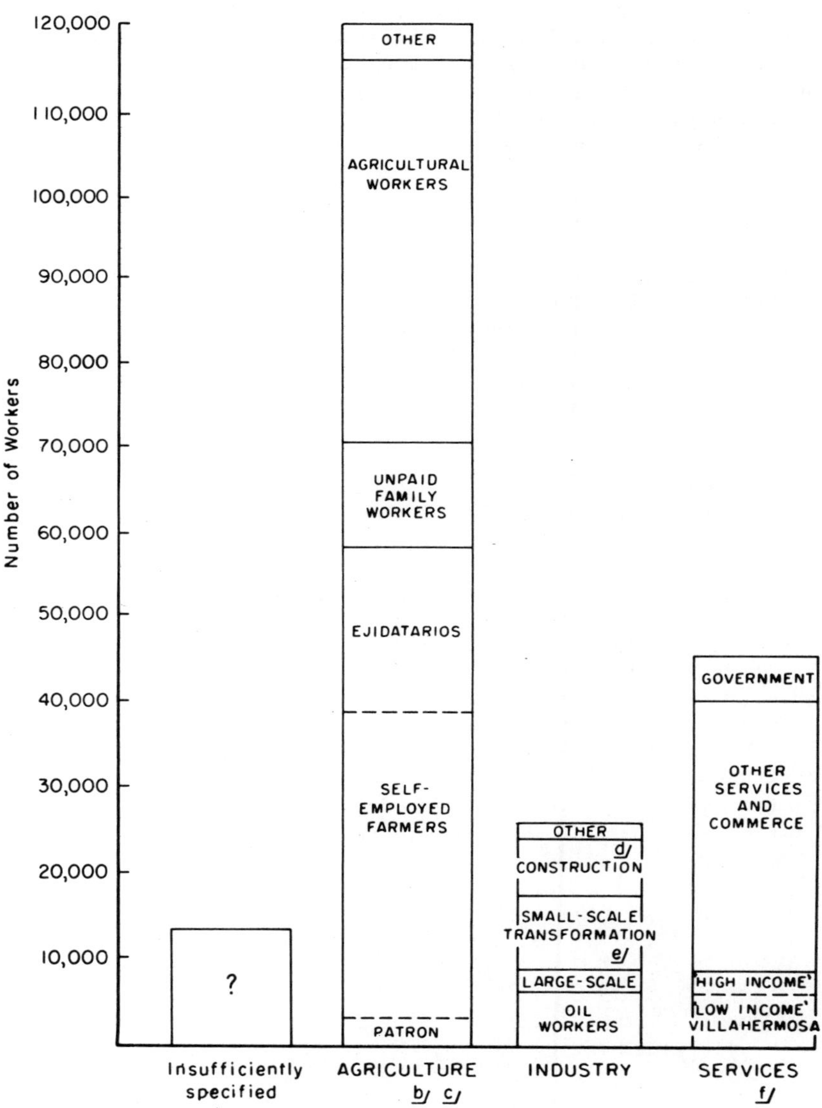

[a]As data on income, branch of work, position, etc., are presented separately, the statistics given in the census do not provide a very clear picture of the working situation of individuals or groups of individuals. This chart is my attempt to combine these variables to summarize what is known about the various groups. It is by no means exact.

[b]Note that in the agricultural sector overall, 14.9 percent of workers (17,208) had no income at all; 81.6 percent (93,943) earned under $1,000 per month (of whom 17.8 percent, or 10,833 earned $500-1,000); 3.5 percent (4,996 individuals) earned over $1,000.

[c]The farmers, including propietarios and ejidatarios, can be differentiated as follows:

Percent of EAP	*Number of Individuals*	*Type*	*Percent of All Farmers*
6	12,000	Subsistence	20.7
17	34,000	Semicommercial	58.6
2	4,000	Collective farmers	6.9
2	4,000	Medium-sized ranchers	6.9
2	4,000	Largeholders with over 50 hectares	6.9

[d]Note that 68 percent of all construction workers were registered as earning under $1,000 per month.

[e]All large-scale transformation workers and 87 percent of small-scale transformation workers earned over $1,000 per month.

[f]There was no way to differentiate income levels for service workers outside of Villahermosa. Within Villahermosa, "high income" workers earned over $1,000 per month, and "low income" workers under $1,000 per month.

Source: México Secretaría de Industria y Comercio (SIC), Censo de Poblasión, 1970.

extremely common for farmers to leave farming temporarily for a period of weeks, months, or years to earn wages off-farm. This was the case for perhaps a fourth of Tabascan farmers during the oil boom.

Of all the rural landless--those who in 1970 were unpaid or paid family workers and those with no ties to the land--I estimated that a third sought some part-time or permanent off-farm employment during the 1970s. In the immediate vicinities of oil production, this ratio was certainly much higher. Counting farmers and landless workers, roughly 30 percent of the 1970 agricultural labor force had participated in some nonfarm petroleum-boom related employment at some time during the subsequent decade. See Chart 2.3.

Immigration and Settlement Patterns

The demand for skilled labor for petroleum development and the general attraction of the economy of Tabasco led to considerable new immigration to Tabasco in the 1970s, though not nearly so much as was frequently claimed at the time (Allub 1983). Out-migration of natives from Tabasco continued to be strong, and most PEMEX-associated workers came to Tabasco only to finish a particular job. Population growth in the 1970s was, in fact, slower than during the 1960s (Negrete 1984, p. 89).

A PEMEX official of the Cunduacán area estimated that in 1980 perhaps 20 percent of the transitorios and a little under ten percent of the permanent workers were Tabascan. He guessed that a fifth of the people then working for PEMEX had also been working for the company in 1970, while the rest were people who were relatives or recommended by former PEMEX workers in Veracruz. New workers after 1978 tended to be Tabascans.

During the decade of the 1970s, Villahermosa's population grew from 99,565 to 107,450 and Cárdenas' from 15,643 to 31,005 according to the official census statistics. The urban population had peaked in 1978-79 as large numbers of temporary workers swelled the population of those cities to an estimated 300,000 and 100,000, respectively. The urban population accounted for a third of the population and Villahermosa alone for nearly a quarter, whereas in 1970 the population had still been predominantly rural, with only a quarter residing in cities of over 10,000 (CNEP 1970-79; SIC 1950, 1963, 1972, 1982).

Of all migrants to Tabascan cities, at least half were from out of state. Whereas in the 1960s there was considerable migration to rural areas due to frontier settlement and the Chontalpa Plan, during the 1970s there was little migration to agricultural areas, with the exception of the Balancán-

Chart 2.3 Estimated Impact of Petroleum Boom on Tabascan Farmers and Farmworkers

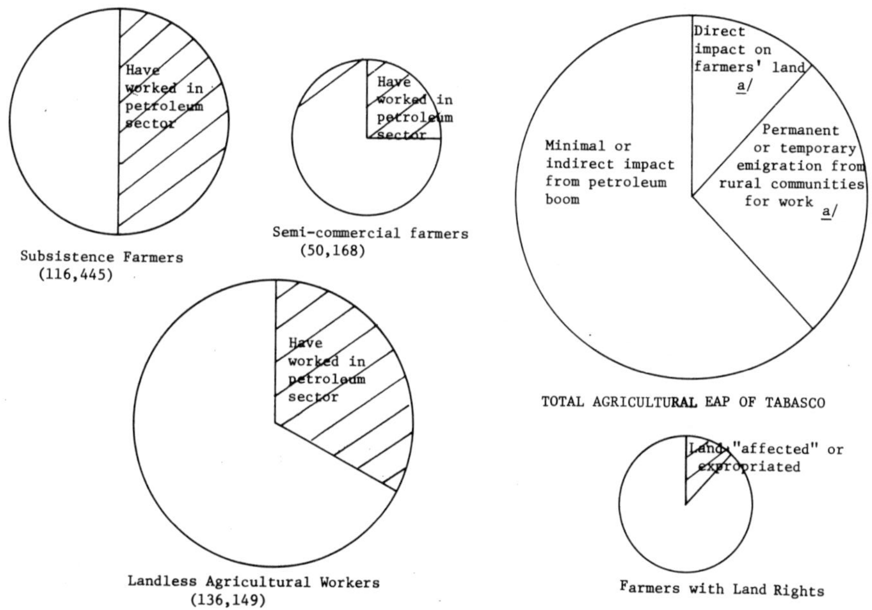

[a]There is probably some overlap between these two groups.

[b]Note that 30 percent of the entire agricultural labor force seems to have participated in petroleum-related employment at some time during the decade.

Source: Based on state expropriation information, extrapolations from the Cacao Farm Case Studies, Malaria Control (CNEP) population information, Population and Agricultural Census information, and the Sociodemographic Survey of Tabasco.

Tenosique Plan area in the eastern panhandle (CNEP 1970-79; SIC 1953, 1963, 1972, 1982).

Pollution

Ecological impacts of PEMEX included crops and pasture contamination, fisheries destruction, death and illness of cattle, water pollution, human eye and lung ailments from air pollution, corrosion of fence wire and zinc roofs, leaks of chemicals from the "holding ponds" by drilling sites, and ruined roads from excessive transit by heavy vehicles (COPRODET 1979b). Some Grijalva Commission engineers were predicting large-scale ecological damage to coastal lakes from the port activities on the coast and from road construction planned by PEMEX, which would change floodplain drainage patterns and increase flood risks.

PEMEX did regularize many operations in the late 1970s and improve pollution controls. In September 1977, an Office of Environmental Protection was set up in Villahermosa by the state government to keep check on progress. While petroleum-related pollution continues to be a serious concern, Toledo suggested that its scale was dwarfed by the ecological impact of massive jungle-felling and fertilizer contamination in the region (Toledo 1982).

Expropriation and "Afectaciones"

By 1980, PEMEX had drilled wells, built roads, pipelines, etc., on the properties of several thousand ejidatarios and propietarios--perhaps five to ten percent of all the state's farms. A small percentage of land was expropriated for permanent activities, but most was "affected" temporarily, as a passage for equipment, for example. Expropriations have been greater on ejidal land than private because, in such cases, the company does not have to pay the value of the land, only improvements. The most serious land use change was the felling of thousands of hectares of coconut plantations near Paráiso for the pipeline and port.

For many farmers, the loss of some land or cattle, or the ruin of annual or particularly perennial crops, could be a severe economic set-back. Compensation for damages was made according to a formula widely felt to be obsolete. In 1977 ruined maize crops were compensated with \$3,500 per hectare versus \$4,090 cost of production; cacao with \$17,000 versus \$30,000 per hectare; and a one-year-old cacao sapling with \$50 vs. \$200 production cost.

Due to farmer outrage, a special claims commission was set up in Villahermosa with representatives from SARH, the

Ministry of Patrimony and Industrial Promotion (SPFI), the Agrarian Reform Ministry, the state government, PEMEX, and the League of Agrarian Communities. Between 1976 and 1978, the value per year of payments for pollution and affectation proceedings grew nearly fourfold; PEMEX had paid over $50 million in compensation by early 1978 (Allub 1980, pp. 44-46).

The fears and insecurities of these farmers had an effect beyond their numbers--in decisions by other farmers not to invest in their land for fear it would be expropriated or ruined.

The Social Impact of PEMEX

The increase in government funds for social infrastructure that accompanied the petroleum boom did indeed have an important positive effect on living standards in Tabasco. Between 1970 and 1980, for example, there was a sharp increase of piped water and sewerage facilities, and near total coverage schooling for school-age children.

But the social fabric of Tabascan society had been rent. Throughout the petroleum areas one heard complaints of violent crime, drinking, pornography. Recreational activities were oriented to young single men with cash, not families. Relations inside families were changing--women were put in charge of farms while husbands and sons left all week to work in Cárdenas; many families had been abandoned.

The active work day for those employed in petroleum activities was longer than in farming. There were long commutes in crowded, dangerous trucks, adding to the anxieties of industrial or construction accidents. Diets were changing; cash needs were increasing with inflation. Even where large monetary indemnifications had been paid for affected land, lack of experience with cash investing had led to tragic squandering of resources.

There had been few efforts by natives to welcome or socially incorporate newcomers, who were expected to leave within a few years. The newcomers' own behavior--families left in Mexico City, weekends and vacations spent away--reinforced the natives' attitude.

"PEMEX has brought to Tabasco three things: lots of filth, lots of undesirable people, and lots of inconvenience" was how one native summed it up. More telling still of the deep resentments felt by many toward PEMEX was the emergence of a horrifying myth. There had long been widespread superstition in Mexico that into all large bridges some child must be thrown, as a sacrifice, to ensure that it stood strong. Supposedly the government stole children to provide this protection. The story had been changed, and I heard versions of it among frightened Mexico City housewives as well as in Teapa, Paraíso, and Cunduacán. It was now PEMEX that stole the children--to throw down the oil wells to ensure that the black gold flowed.

Social and Political Conflict with PEMEX

The haste with which early PEMEX activities were carried out created lasting resentments in Tabasco: drilling rigs were built on farmland before expropriation papers were ever finished; payments were delayed for over a year. Roads were destroyed by overuse.

External control over PEMEX activities was sharply eroded by the 1978 reform of Articles VII and X of Article 27 of the Mexican constitution. This made it impossible to refuse PEMEX occupation or expropriation and gave petroleum production priority over any alternative land use (Tirado 1978, p. 33).

PEMEX at one point proposed siting of the new petrochemical plant in a densely inhabited and highly productive cacao zone of Cunduacán. It was eventually moved to Gil Perez in Centro, after demonstrations and protests by cacaoteros. But in a fit of pique

> *PEMEX announced that it would move all its new installations or a large part of them to Chiapas, in view of the fact that in Tabasco they were not well-received. This was fundamentally an attempt to pressure Tabasco to accept PEMEX unconditionally. And to make that pressure stronger, PEMEX halted the construction of housing in Villahermosa [Tirado 1978, p. 103, my translation].*

On another occasion, the army was brought in to force farmers protesting lack of indemnification payments to allow PEMEX trucks to enter their ranchería. In January 1980, an estimated 10,000 armed campesinos cut off roads in the Chiapas petroleum zone, in protest of water pollution that had supposedly destroyed much of their cacao crop ("Armed" 1980).

The League of Agrarian Communities organized some 3,000 individuals in 1980 to improve campesino political organization. The "Pacto Ribereño," a group of 7,500 campesinos, organized legal help to defend their rights and payments of indemnifications (Tirado 1978, p. 43).

Temporary workers for major PEMEX contractors also periodically protested conditions. Attempts in 1974 to organize a union and require transport services and the usual benefits of Mexican working standards led to the firing of 800 workers.

The workers' reaction was so violent that 1,000 soldiers were called in. Shortly after, a general strike was called. It lasted a month and ended with a contract. But in the next few months, the companies claimed to finish work, closed contracts--and then began work again with different people. Other strikes followed, but actions by the PEMEX union and other groups to break up organization were successful. In 1979-80 there were no guarantees at all for transitorios.

Collaboration between the state government, the farm producer associations, and supporters in Mexico City forced some restrictions on PEMEX, including an obligation to inform the Tabascans in advance of proposed activities. But the collaboration was shaky. The general population resented that the main beneficiaries of the boom were the "'rentistes,' contractors, large merchants, speculators, landlords, bankers, industrialists, PEMEX union officials and privileged PEMEX workers" (Tirado 1978, p. 79)--and suspected the government of collaborating with them.

THE CONTRADICTION: AN AGRICULTURAL BOOM

Despite all of these serious problems with petroleum development, the agricultural sector over the same period experienced its own "boom" illustrated in Chart 2.4. While livestock production figures from different sources varied significantly, all showed a rate of increase in the 1970s equal to or greater than the increase during the 1960s (SARH; DGEA).

Cacao production rose dramatically during the 1970s, with historically high production even in 1982. Low production in 1981, and probably that of 1983, was due to weather. Coconut production was high, though stagnant 1975-79. Production in 1979-82 reached record levels. Banana production rose sharply throughout the 1970s, declining slightly in the early 1980s. Sugar cane production, which rose sharply until 1976, plateaued through 1979 then entered a period of decline through 1982.

Basic grains production was not so impressive. After a brief period 1975-76 of record high harvests, production of rice declined sharply, only to recover 1981-83. Possibly sugar cane was being substituted for rice on the collective ejidos of the Chontalpa during 1981-83. Maize production was stagnant, though production levels in the 1970s were significantly higher than those in the 1960s. Due to excessive rains 1979 was a very poor year. There was some evidence that maize farmers (about 15 percent of producers) emigrated during the 1970s in large numbers for off-farm work; whether temporary or permanent was not known. Bean production, while trending slightly upward 1973-79, declined thereafter with the exception of 1981-82.

Chart 2.4 Agricultural Production in Tabasco, 1960-83[a]

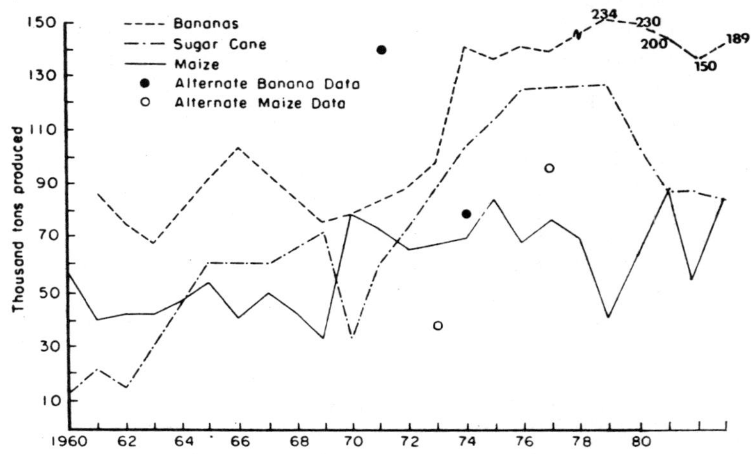

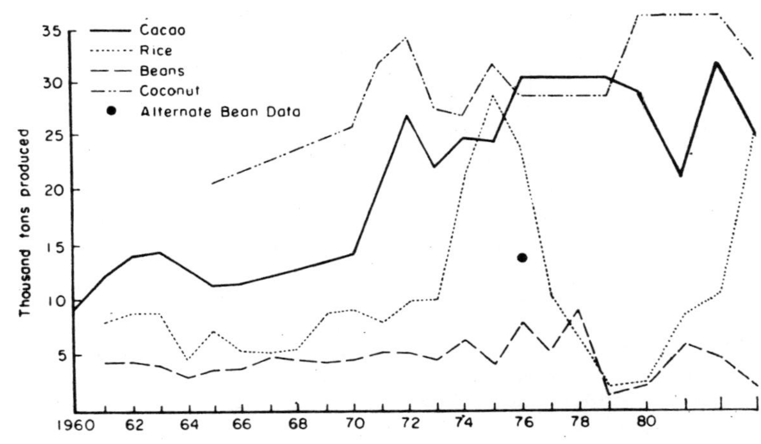

[a]Data for sugar cane production have been charted in units of ten thousand tons produced.

Sources: Secretaría de Industria y Comercio, Dirección General de Economía, *Censo Agrícola, Ganadera y Ejidal*, 1960 and 1970, "Tabasco," Cuadro 10, Mexico, D.F.: Eduardo Bolís Villanueva, *La Economía del Estado de Tabasco*, Colección de Estudios Económicos Regionales, Sistema Bancos de Comercio, Mexico, D.F., March 1976; Manlio Tirado, *Tabasco y el Gasoducto* (Ediciones Quinto Sol, S.A.: Mexico, D.F.), 1978, pp. 54-60; SARH, Departmento de Economía Agrícola, Residencia Tabasco, Villahermosa, Tabasco, statistical records; INEGI and COPLADET, Anuario Estadistica de Tabasco 1984, INEGI: Mexico, D.F.

Though not shown, black pepper and vegetable production in Tabasco declined between the mid-1970s and 1983. Tree fruit production rose in the early part of the 1970s, remained stable, then declined slightly in the early 1980s (INEGI/COPLADET 1984).

The surprisingly successful performance of the sector can be attributed to four major factors: rising agricultural prices, new agroprocessing investments, the state program of assistance to and investment in agriculture, and the flexibility of the smallholder farm sector.

Agricultural Prices

Chart 3.5 shows changing prices for the major commercial products. Price levels for cacao, copra, beans, and beef were particularly favorable; those for sugar cane and bananas rose in real terms. For the export crops, devaluation of the peso in 1976 helped to spur price increases and production. In the context of an inflated cost of living, the contrasting relatively low price of maize and rice seems to have justified grain farmers' negative response.

Agroindustry Development

Value-added from agriculture was further multiplied by substantial growth in the agroindustrial plant of Tabasco during the past decade. New investments included expansion of the coconut oil factory, a new modern slaughterhouse for cattle, a transport center for cacao and five cacao fermenting plants, a banana warehouse, and a pepper-drying plant--all owned and run by their respective farm producer associations.

Two privately-owned cheese factories and a poultry slaughterhouse also went into operation (Gob. Tabasco 1979b, p. 52), and two new cane mills were built by the federal government in Hermengildo Galeana (1971) and Benito Juárez (1975), nearly doubling state milling capacity (CNIA 1976, pp. 30-33).

The Tabasco State Development Plan

Although agricultural prices and new industry presented the promise of agricultural prosperity, it was strong federal and state support for the sector that generated the necessary investment confidence, political support in conflicts with PEMEX, and generation of rural employment and investment in zones outside petroleum-drilling areas.

Native Tabascans outraged at PEMEX disruptions began a national campaign in the mid-1970s to bring PEMEX under some

Chart 2.5 Prices for Major Agricultural Products of Tabasco, 1970-79

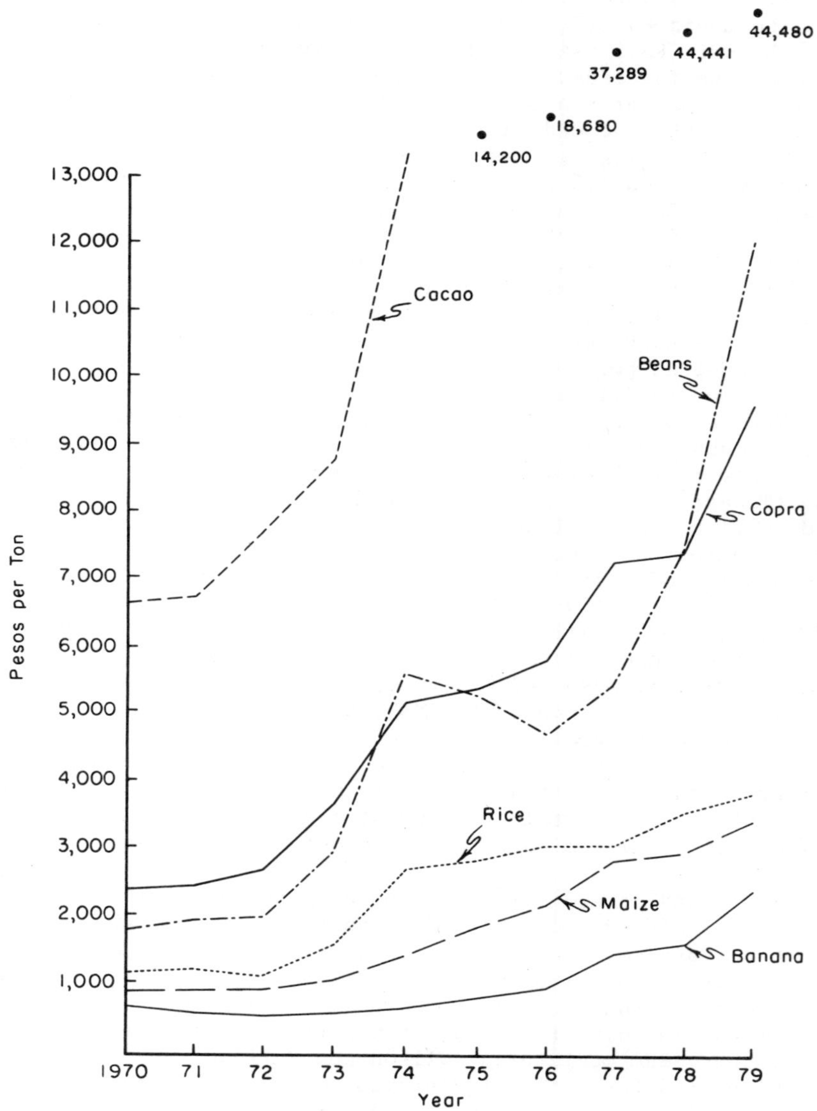

Source: Mexico, SARH, *Econotécnia*, Vol. IV, No. 8, August 1980, pp. 18-30.

control. In the midst of the heated national political debate about the wisdom of breakneck petroleum development, they received a sympathetic hearing.

The López Portillo government, inaugurated in 1976, chose Ing. Leandro Rovirosa Wade as state governor, whose family had been active for decades in Tabascan agriculture and politics. As former head of the powerful Water Resources Ministry (SRH), he was a symbol of dedication to agricultural development.

With the collaboration of the new Programming and Planning ministry, the Comité Promotor de Desarrollo del Estado de Tabasco (COPRODET) was set up, intimately connected with the state executive branch. Tabasco received more professional, logistical, and political support for state planning than did any other Mexican state (Wilson Salinas 1983). PEMEX was ordered to collaborate with COPRODET and the Governor to resolve issues of mutual concern, such as materials bottlenecks and transport difficulties, and to inform the agency of its planned activities.

The Tabasco State Plan which was developed envisioned a future Tabascan economy oriented around four axes: petroleum, hopefully petrochemicals; agriculture and agroindustry; fishing (Tabasco was the key state in the National Fishing Plan); and tourism. Agriculture, especially, and tourism would be the principal employers in the state and would help to right the skewing of state income distribution expected from petroleum development (COPRODET 1979a, p. 54; Scherr and Scherr 1981). The plan called for investment per capita to be highest in the nonpetroleum municipios, and in urban centers outside Villahermosa and Cárdenas.

To offset the impact of PEMEX, the program would train skilled workers locally to reduce immigration; bring sports centers and concerts to offset the influence of petroleum workers' cantinas; attract tourists so that petroleum workers were not the only outsiders; and strengthen agriculture to offset petroleum dominance of the economy (Scherr and Scherr 1981, p. 12).

The focus of economic development plans was the already densely populated area from Cárdenas to Villahermosa, and north to Paraíso. While these three cities, as the principal urban centers, received a disproportionate share of government investment during the 1970s, there was also significant investment in communications, transport, and industry sectors in nonpetroleum municipios (Wilson and Salinas 1983, p. 161). Petroleum led the sectoral distribution of the budget (42 percent)--followed by industry (19 percent), housing and related infrastructure (11 percent), agriculture (11 percent), and tourism (9 percent).

The most important investments during the 1976-82 period were the Gil Perez petrochemical plant and the integrated petroleum port in Dos Bocas. A modern new airport for Villahermosa was completed in 1979, and work began on a four-lane highway from Coatzacoalcos to Villahermosa. Also in the capital were "Tabasco 2000," a large-scale urbanization project, and a convention-museum-tourism center. The "Industrial City" of Villahermosa was constructed in 1972 and began operating in March 1973. Under the "Alliance for Production" an agreement was signed between the governor and private industrialists for investments there. Over 6,000 jobs were expected from the 83 projects under that agreement. By 1978, 28 industries had contracted with the City, of which nearly a third were agroprocessors or storage facilities for agricultural products (Bolís 1976, p. 46; Gob. Tabasco 1980b, p. 46).

Of $8,500 million pesos for Alliance expenditures in 1979, nearly a quarter went to agriculture (mostly production credit), 37 percent to industry (mostly construction materials), 12 percent for human settlements, 21 percent for tourism (COPRODET 1979a, p. 47). Including PEMEX and other public and private expenditures, total Plan investments in 1979 ran to over $30,000 million (or U.S.$1.3 billion dollars). Of this, nearly 40 percent were to have been made by private interests, in return for tax advantages, credit assistance, and infrastructure provision.

Of State Plan funding, 54 percent came from the federal government, 30 percent from federal financing for private investment projects, 10 percent from private sector financing, and six percent from state and local financing. The federal sources included a substantial amount of international financing for PEMEX investments, urban infrastructure, and agriculture.

Government Investment in Agriculture

Although in relative terms agriculture represented only 11 percent of the budget, in real terms this was an unprecedented level of assistance to the sector. Map 2.6 shows the major agricultural projects in Tabasco in 1980. These included the rainfed agriculture district programs, the Plan Chontalpa, Plan Balancán-Tenosique, and the Grijalva Commission drainage projects.

District I emphasized livestock production; District II, agricultural production. SARH ran a series of programs including input production; research; technical assistance; crop improvement programs for cacao, coconut, and bananas; promotion and planting programs for rubber, cassava, cashew, achiote, and jamaica; and programs for fruit trees, apiculture, and basic crops (COPRODET 1979d, p. 49).

The Chontalpa Plan was reorganized, and in the late 1970s it showed much improved statistics on production, credit repay-

Map 2.6 Agricultural Development Projects in Tabasco, Mexico

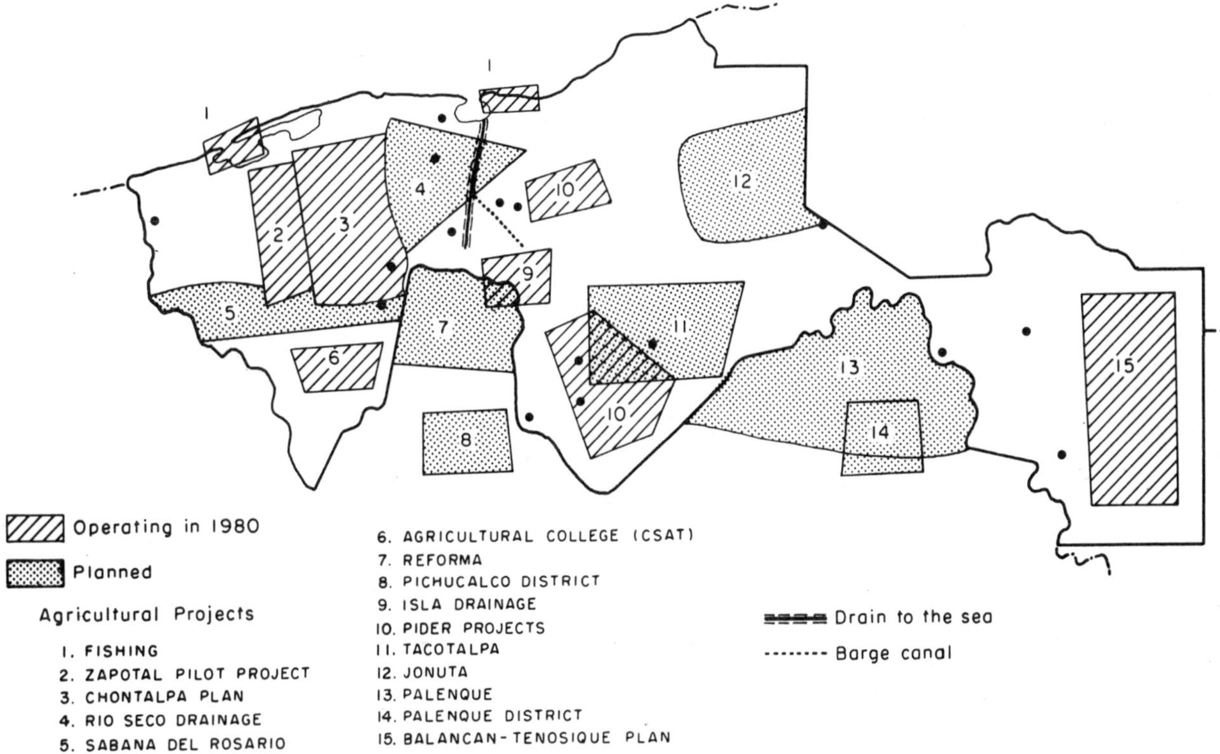

Source: Mexico, COPRODET, "Plan Sectorial: Agropecuario y Forestal," *Plan Estatal de Desarrollo*, 1979.

ment, and labor absorption. The Balancán-Tenosique Plan--another collective enterprise primarily for cattle production --was begun in 1971 on 61,000 hectares. In the early 1980s, it was the site of an intensive rice production project.

The great "Drain to the Sea" was proposed and initiated to protect the Central Chontalpa area from flooding and to open new land for cultivation. This costly project, which would extend from Centro to Paráiso municipios, was highly controversial for economic, political, and ecological reasons (Toledo 1982).

Other drainage projects were underway in Zapotal (a PRODERITH pilot project), Isla, and Río Seco. Originally the Chontalpa Plan's second phase, the Río Seco project, covered 37,000 hectares in the most densely populated part of the state and required completion of the "Drain to the Sea" to be successful. There were a number of other projects to build drains, dikes, and bridges, and to desilt drains and rivers, mainly in the Chontalpa region.

The federal Intergrated Program for Rural Development (PIDER) operated in 1980 on nearly eighty ejidos, with over 4,500 ejidatarios, mainly in the Sierra and Old Chontalpa regions. Its projects included drains, roads, fencing, and plantation establishment. The Planning Commission for Marginal Areas (COPLAMAR) worked in agriculture and social infrastructure with indigenous groups, mainly in the semi-flooded Old Chontalpa areas. Storage capacity for CONASUPO in the state was increased from 1,500 tons in 1977 to 21,500 tons in 1979.

In 1979, total agricultural investments amounted to 14 percent of that year's estimated total agricultural production. In addition, federal agricultural credit guarantees (FIRA) amounted to well over half a billion pesos, mostly for long-term livestock loans (Gob. Tabasco 1979b, p. 47).

The 1982 economic crisis had important effects on Tabasco, but by and large the state was better off than most other regions of Mexico. A reduction in PEMEX construction was probably more positive than negative for the local population, given continuing high levels of income accruing to the state from existing oil facilities. Devaluation offset in part the effects of declining international commodity prices, and export agriculture in Tabasco continued to do relatively well. Maize production continued to decline, however, despite slight increases 1981-82 due to temporary SAM-related policies. The new governor Enrique González Pedrero (1982-) directed substantially more resources to the agricultural sector and rural areas than did Rovirosa, and most of the major agricultural investment and infrastructure projects were being completed as scheduled (Grijalva Commission, senior engineers, personal communication, January 1985).

THE CAMPESINO ECONOMY AS THE PRINCIPAL STABILIZING FACTOR IN AGRICULTURAL PRODUCTION

While good agricultural prices and strong governmental support were critical to Tabasco's agricultural "boom," they were not sufficient in themselves. Upon closer examination of the production statistics and after widespread interviewing among farmers, it was evident that large numbers of farmers did respond to the disruptions of the petroleum boom by reducing production. These were the larger farmers who depended upon hired labor for farming. Even when agricultural prices were such as to permit farm profits, these producers frequently chose to invest their resources in livestock, tertiary activities, or urban real estate speculation.

The stability of agricultural production in Tabasco came mainly from the campesino sector, in particular the farms small enough to be operated predominantly with family labor, yet of sufficient size to support a family. Their rationale for production decisions was based on considerations best explained by the theory of "family economy." These farm families both supplied labor on the off-farm labor market and hired labor at critical times on their own farms. Their supply of labor was flexible, so that when high farm prices and high off-farm wages warranted it, they could expand employment both on- and off-farm.

The peak labor demand periods in Tabasco were March-May and November-December. By contrast, the main period for construction activity on roads, drains, and other projects was February-May, the dry season. This labor seasonality led to differing composition and size of the agricultural labor force at different times of the year (wreaking havoc on macrostatistical analyses of employment in the state).

To meet labor peaks, a large part of the smallholder sector (nearly 60 percent of farmers) worked periodically for largeholders, and often for each other, in between work requirements on their own land. Such work relations were often formalized, with largeholders providing credit and help during sickness in return for valuable peak season help, and with neighbors joining in godparenthood and work teams.

Many individuals in the Census "farmworker" category were in fact sons of smallholders who were paid by parents or neighbors, and probably lived at home, hence not "landless" in the strict sense. Many others were designated "ejidatarios without land rights." This status, though often literally the case, frequently masked a cooperative situation in which, for example, several brothers farmed together, but only one was represented on the ejidal council or on official records.

Permanent hired labor was uncommon, but probably over 50 percent of all farmers--and this included many with under five hectares of land--hired nonfamily labor at some time during the year. The level of demand for such work depended on product price and yields. When prices were high, it was worthwhile for farmers to hire help for more intensive cultivation. When yields were high, either due to natural conditions or to more intensive cultivation, much more hired labor was required at the peak harvest seasons.

Potential family labor not usually employed in agriculture provided much of the flexibility. Appendix C compares data on agricultural labor from the Population and Agricultural Census. The marked differences in employment figures between the two clearly illustrate the critical role and extent of agricultural labor by family members who do not consider themselves to be principally involved in the agricultural sector. Such workers include farmers' wives and daughters, children usually in school, and children or cousins with other jobs who help on the weekend and afternoons on the relative's plot at peak labor times. Indeed, Tabasco's ability to triple banana production and cane production, double cacao production, and increase coconut production by 50 percent, while reducing maize production by only 16 percent, co-ld not have taken place in the middle of an off-farm employment boom and when technological changes were only marginal, without a major net increase in the absolute number of agricultural man-hours worked.

In Part, this was made possible because the major petroleum zone was so densely populated that good communications with urban centers permitted farm family members to work for PEMEX or the constructoras without leaving home. In the Chontalpa, even farmers who were away all week returned to do farm work on the weekends. Hence, rural emigration rates were relatively low, despite extensive employment off the farm.

More important, the farm budgets of these family-based farms were organized to minimize the impact from petroleum-induced economic changes. Not only was hired labor a relatively small part of the budget, but the families could sometimes substitute family labor for purchased inputs that became too expensive (e.g., substituting hand labor for herbicides).

THE CACAO ECONOMY AS A CASE STUDY

The importance of the mixed subsistence-commercial smallholders in offsetting the economic disruptions of petroleum development was clearly illustrated in the case of the Tabascan cacao farmers. The cacao crop required extremely high labor inputs, and farmers would be expected to be particularly sensitive to off-farm wage changes. Yet production rose, area planted expanded, and cacao production remained Tabasco's most important economic sector in terms of rural income and employment.

3

The Cacao Economy of Tabasco

Junto a una taza de té, platicamos; junto a una de café, discutimos; con un chocolate, ninos, mozos y viejos, sentimos el placer de la vida.

Beside a cup of tea, we chat; beside one of coffee, we argue; with a cup of chocolate we all--children, adults and older folk--feel the pleasure of life [quoted in Padilla 1971, my translation].

--*Enrique Perez Gutierrez*

Cacao--delicacy of the gods

--*Linnaeus*

Cacao has for the inhabitants of Tabasco an importance beyond the economic: It is one of the identifying features of Tabascan culture. But from a staple component of traditional subsistence livelihood, its production has evolved into a dynamic part of the state's modern economy with sophisticated processing and marketing components and a structure of production quite unlike that of even a few decades ago.

AN INTRODUCTION TO CACAO

Cacao is a tree naturally found in the lower stories of the tropical rain forest. From its seeds chocolate is made.

Hernando Cortéz, after discovering cacao in the Conquest of Mexico, brought it to Europe. When the Europeans learned to combine the bitter paste with sugar, a popular drink was born, creating a commercial demand for cacao. In the nineteenth century chocolate industries developed in Europe.

The principal trade products of cacao are the cocoa bean, typically fermented; cocoa butter, the fatty part of the seed which has numerous food and nonfood uses; cocoa powder and cakes used for cooking; cocoa paste; and of course the more elaborated chocolates for drinking or eating.

In 1978, 28 percent of world cacao production came from South America, half of which was from Brazil. Seven percent was from Central America; 60 percent from Africa (Ghana, Nigeria, Ivory Coast, and Cameroon), and five percent from Asia and Oceania (CONADECA 1976, pp. 2-6). The major consumers were in North America and Europe.

Technical Processes of Cacao Production

Cacao (*Theobroma cacao*) in the wild is a small tree in the lowest story of a rainforest, about six to eight meters high. Optimum conditions for its production require temperatures of 21-32 degrees Celsius; over 1,250 millimeters rainfall with no month under 125 millimeters; altitude under 300 meters; and well-drained and aerated, deep soils with a crumb structure (Purseglove 1968, p. 575). Established cacao is not hurt by flooding for several weeks (Urquhart 1955, pp. 34-35).

Properly raised and pruned, cacao develops a main stem one to two meters high, at which point it divides into a "fan" or "jorquette" of three to five horizontal limbs. A sucker (or "chupón") immediately below the jorquette on the main stem leads up to a new jorquette. Buds on the stem and leaves are called "cushions," where the flowers arise. After the flower forms, a "cherelle" (or small pod) emerges; only five percent of these are pollinated (Stavins 1971, p. 6). The cherelles attain full size four to five months after sterilization and ripen in a month. Each contains 20 to 60 seeds. Harvest can take place over two to three weeks, and once harvested, the pods keep for up to a week (Purseglove 1968, pp. 578, 589). In general, the farmer harvests each tree every fifteen days during peak production season and once a month at other times.

There are three main types of cacao: the criollo, the forastero, and their cross. The first produces the highest quality bean, but takes six years to come into production, is susceptible to disease, and produces only a quarter kilo of dry beans per tree. The forastero is more resistant, very productive, and begins producing in the third year. The clonal hybrid and "Ceylán" (or "grafted") crosses are very tall and somewhat more productive, on average, than the forastero.

The color of the criollo bean is white; the forastero, deep purple; their cross, a range between the two. The criollo has a higher fat content and pronounced chocolate flavor after fermentation and is very aromatic. These characteristics are minimal in the forastero and limited in the crosses (Bracho Roblín et al. 1972, Chart 9). Mexico began substituting its traditional criollo plantations with forastero and Ceylán varieties in the 1940s.

Shade control is important in cacao management. Seedlings must be grown under shade, both lateral and overhead. After the second year, the lateral shade can be removed and only the main shade trees kept. The desirable level of shade in adult trees is determined by the fertility of the soil. There is an inverse relation between them such that extremely well-fertilized soils need little shade.

Pruning is done at the end of the dry season (Braudeau 1970, pp. 176-79). In Mexico the various prunings are differentiated into the "desmadre" (cutting of the "mother" trunk to leave a stronger principal branch); the "desmamone" (cutting of young shoots); the "deschupone" (cutting off the suckers); and the "poda" (the first formation cutting).

The process of fermentation removes mucilage from around the beans, diffuses the purple pigment of raw beans, and removes astringency. Subsequent drying should decrease moisture content from 56 to six percent for storage. About 750-1,000 cured beans make a kilo (Purseglove 1968, p. 589).

Cacao plants remain productive for decades. They reach maximum productivity around twenty years of age and decrease production after forty. Other factors leading to declining productivity are a poor canopy, weeds, loss of leaves through disease, poor or too dense shade, and parasites (Braudeau 1970, p. 215). In Tabasco the principal diseases and pests of cacao were: black rot (*Phytophthera palmivora*), Ceratostomelia, thrips (*Selenothrips rubrocinctus*), aphids, and ants. There was also a relatively new threat of the fungal disease "frosty pod rot" (*Moniliaphthora roreri*) which had spread from South to Central America. Insecticides were used to control thrips, and Bordeaux and other copper-based solutions were used to combat fungal infections.

Plantations are renovated by one of three methods: regeneration, rehabilitation, and complete replanting. The first method requires planting seedlings under existing trees and removing the latter by stages. The second interplants young trees by sections, every year, then requires pruning existing trees to decrease their yield gradually in favor of the saplings. In complete replanting, old trees are cut down and new plantations begun from scratch; the new stand is more productive, but the cost is high (Kolawole 1972, p. 161).

In Tabasco, the average area in cacao is 2.7 hectares. The situation is similar in most producing countries. Of the six leading cacao producers, in Cameroon, the average is 1.2-2.0 hectares; in Ghana, two hectares; in Nigeria 1.4 hectares. Only in Brazil have large-scale plantations in this century been successful: over 80 percent of land planted in cacao is in holdings of over 30 hectares ("Consulta" 1977, pp. 9-31).

Labor Use in Cacao Production

The labor input in cacao production is considerable; even more so in frontier areas where new plantations must be cleared. Chart 3.1 illustrates the recommended calendar of cacao production activities for Tabasco. November, December, and January are the prime months of the "Cosecha" (Harvest). A second peak occurs in May (the "Alegrón"). Year-to-year variability in both major harvest and minor harvests is extreme, as shown in Chart 3.2. During the peak month of cacao harvest, production varied from under 4.5 million kilos (1977-78) to 8.5 million kilos (1978-79).

Labor use is a function of intensity of input applications, weeding and pruning care, requirements for drainage systems, and, particularly, harvested value of the beans, which determines time spent on harvesting and the financial incentive to apply inputs on the plantation. Post-harvest processing such as pod-breaking, bean-washing and drying, or fermentation account for a large part of all labor costs. Reported labor use in Africa, the West Indies, and Mexico ranged from 27 to 130 man-days per hectare per year, with wide yearly differences due to establishment procedures and yearly harvest increases in the development phase (Berry 1975; Bracho Roblín et al. 1972; Braudeau 1970; Hill 1970; SARH 1980a; Olayemi 1971; Urquhart 1955).

There is also wide variation among producers in the distribution of labor between cacao activities. This is largely because labor can nearly always be substituted for the capital needs of cacao production. Such common purchases as seed, fertilizer, and pest and disease control chemicals can be substantially replaced by home nurseries, on-farm bean selection, and scaring away or killing pests by family members. There are a few tools required that cannot be hand-made from low or no-cost materials.

Land is the critical production input. However, there are almost no true economies of scale to large plantation size, because of the inappropriateness of mechanization and, in Tabasco, the long history of a central marketing organi-

Chart 3.1 Recommended Calendar of Cacao Production Activities

Month	Activities		
OCT	spray for black rot[a]	new plantations	harvest
NOV			
DEC			
JAN	spray with copper sulfate		
FEB	fertilize soil		
MAR	spray insecticide for thrips[b]		
APR	maintain drains		
MAY	prune	disinfect pruning cuts	
JUN		fertilize resow soil remove old trees	
JUL	insecticide for "pulgon," "salivazo"[c]	resow	new plantations
AUG	pesticide for lizard	remove excess new shoots	
SEP	spray for black rot		

[a]*Phytophthora palmivora Butl.*

[b]*Selenothrips rubrocinctus Giard.*

[c]*Aphis, gossypii Glover; Stephanoderes guatemalensis Hopkins; Xyleborus affinis Eich.*

Source: SARH and Unión Nacional de Productores de Cacao, "Calendario para el Combate de Plagas y Enfermedades de las Plantaciones de Cacao y Labores Culturales," Villahermosa, Tabasco, September 1979.

Chart 3.2 Cacao Production Cycles in Mexico, 1971-79

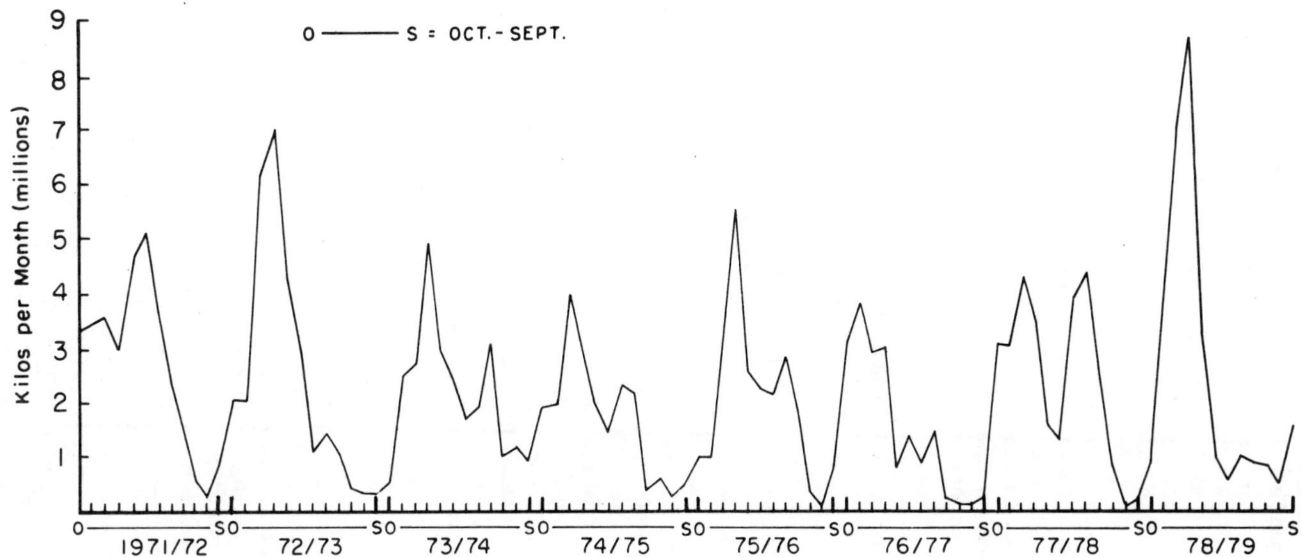

Source: Mexico, Unión Nacional de Productores de Cacao, *Boletín Informativo #4*, June 1979 and *Boletín Informativo #1*, 1981, "Producción Mensual de Cacao por Ciclos Agrícolas.

zation. As Kolawole says for Nigeria, "Yield and not size is the principal source of cost economies" (Kolawole 1972, p. 238).

Yields in Tabasco averaged 538.5 kilograms per hectare in the 1970s, with official yields 1967-78 ranging between 458 and 841 kilos per hectare (UNPC 1977-81; SARH 1978b). Between 1930 and 1971, yields varied from 162 to 439 kilos per hectare or .294 to .798 kilos per tree. The lowest level was in 1930-34; the highest 1962-63 (Padilla 1971, p. 90). This compared favorably with international figures: in Venezuela and Cameroon, average yields were only 270 kilos per hectare; in Nigeria 250-300; Ghana 330 (Purseglove 1968). By 1983, some areas of Tabasco had yields as high as 1,000 kilograms per hectare.

THE STRUCTURE OF THE TABASCAN CACAO ECONOMY IN THE 1970s

Until this century, cacao production in Tabasco took place primarily on individual campesino smallholdings, or on larger plantations using hired or ecomendado labor. Marketing was by private commercial intermediaries, often foreign.

The state had 3.5-5.0 million trees in 1854, of which 79 percent were in the Sierra, mainly Teapa. In 1890, the number was down to three million trees, and the Sierra had 39 percent. By 1950, the state again had five million trees, 74 percent of which were in the Chontalpa.

Cacao production in the early twentieth century amounted to under 1,000 tons per year on under 3,000 harvested hectares. There were probably no more than 1,500 producers. Fifty years later there would be about 16,000 producers on 42,000 hectares, marketing over 30,000 tons per year.

Cacao Zones of Tabasco

Map 3.1 shows the cacao-growing areas of Tabasco. For historical and technical reasons, the state can be divided into five cacao regions: Huimanguillo, the Chontalpa Plan (parts of Huimanguillo and Cárdenas), the Sierra, the Old (eastern) Chontalpa (along the Cunduacán road); and Río Seco (along the Cárdenas-Paráiso road).

Río Seco was by far the most important zone, with half the producers and cultivated area. Its soils were the most fertile and best drained; land was not so subdivided as in the eastern Chontalpa (with the exception of Paráiso). The "old" (eastern) Chontalpa was more subject to flooding problems and in general was less prosperous. In the Chontalpa Plan, cacao production was divided between collective

Map 3.1 Cacao-Growing Zones of Tabasco

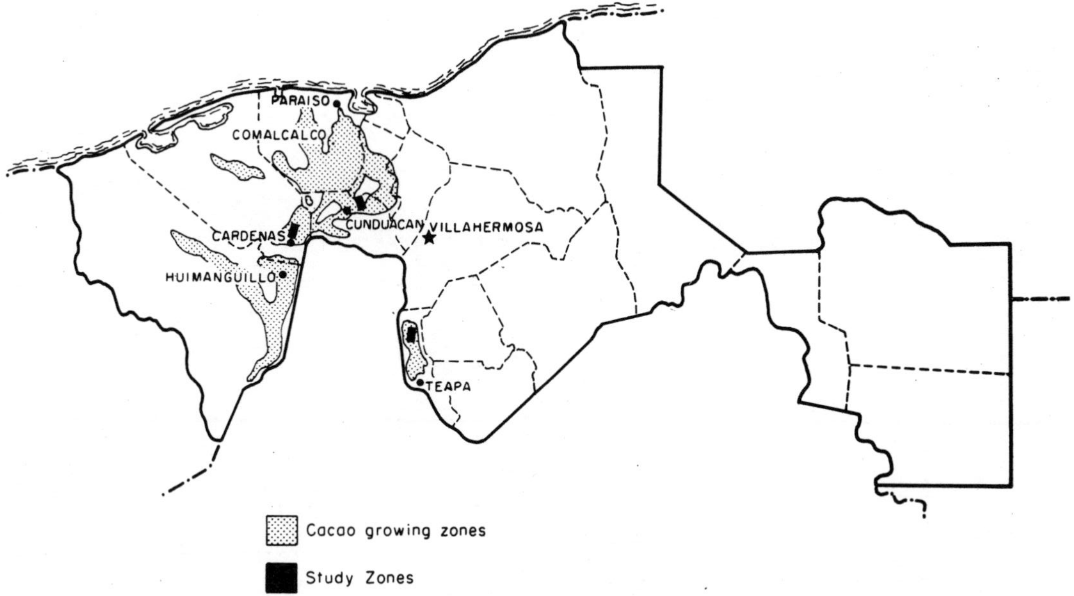

Map 3.1 continued

Note: Area planted estimated from 1978 Cacao Census and from sales data of the Unión Nacional de Productores de Cacao. Producer numbers estimated from 1978 Cacao Census, UNPC estimates and the Promotor Chontalpa. The latter are very rough approximations.

[a]*Cacao Area Planted, by Municipio (hectares)*:

Municipio	Hectares	Municipio	Hectares
Cárdenas	6,032	Jalpa	4,435
Centro	72	Paráiso	4,931
Comalcalco	16,258	Teapa	1,148
Cunduacán	6,556	Tacotalpa	101
Huimanguillo	2,368	TOTAL	41,900

[b]*Cacao Zones, Area Planted and Number of Producers*:

Zone	Area	Producers
Sierra	1,250 hectares	500 producers
Huimanguillo	2,300	500
Plan Chontalpa	1,000	2,000
Eastern Chontalpa	13,250	2,500
Río Seco	18,000	7,500
Coast	5,000	2,000
TOTAL	41,000	15,000

Source: SARH, Representación Estatal, "Tabasco," *Plan Indicativo Agrícola 1978-82*, Villahermosa, Tabasco, December 1977.

and private plots. In the former, plantations were operated on a large scale, on land newly drained in the late 1960s and 1970s. Huimanguillo production was on lower quality soils and tended to be on larger plots of land owned by recent colonists. Sierra growing conditions suffered from higher rainfall, lower temperatures, and poorer soils. Table 3.1 provides details about the regional variation in PEMEX activity, cacao labor scarcity and wages, economic conditions in general and specifically for cacao, and farmer migration patterns.

The history of the cacao regions, particularly the changes brought about by new river routings and farmer migration, were discussed in Chapter 2. They explain the present distribution of farmers and farm size; the New Chontalpa and older nonflooded zones of the Old Chontalpa are heavily populated and dominated by tiny cacao farms, the minifundia. Areas opened up relatively recently in the Chontalpa are less populated and minifundization is less advanced. In Huimanguillo and Teapa, properties tend to be larger because ejidal settlements are relatively new.

Evolution of Cacao Farmholdings

Throughout the centuries of cacao production in Tabasco, the diversified subsistence farm household was predominant. The amount of land planted to cacao was dependent on the availability of family labor, probably averaging four to six hectares. This general pattern was broken by those farmers who were first opening up new lands to cacao production, who generally had temporarily smaller plots, and largeholders specializing in export, with some control over a labor force. With the takeover by Spaniards of indigenous farms in the 1600s, and later the commercial farms in the early twentieth century, very largeholdings developed throughout much of the state dependent on encomendado or debt-peonage labor.

The structure of the cacao economy, however, was considerably different in the 1970s--the result of population growth and the closing of the agricultural frontier. The temporary phase of growing crops for subsistence while awaiting the cacao plantation's coming into production occurred less frequently. The Sociodemographic Survey of Tabasco showed that although 30 percent of farmers planted cacao alone, the rest planted maize, bananas, coconut, or sugar cane, or ran cattle, as well. Farmers with under two hectares were more likely to produce cacao alone or with maize.

The shortage of available land for settlement by the children of cacao farmers has led to accelerated minifundization. Meanwhile, many largeholders moved out of cacao into livestock production, and the average size of the cacao large-

Table 3.1 The Tabasco Cacao Economy--Regional Differences

	PEMEX Activity	*Cacao Labor Scarcity*	*Cacao Labor Wage*
Eastern Chontalpa	Most PEMEX activity: 500 workers in union (half transitorios; half permanent); in 1980--500 workers in PEMEX construction CTM membership (not all PEMEX) 1978--5000; 1980--2500 PEMEX work mainly in dry season	Significant labor scarcity Accelerated substitution of herbicides for machete weeding Most farmers who worked off-farm in 1960s no longer do so because farm incomes have improved. Others switched to off-farm construction	Highest wages--here minimum of \$120 per day for 6-7 hours day Wage on ejidos reported higher than on private farms because cacao income is more essential
Río Seco	Minimal petroleum activity Minimal direct impact of PEMEX on population	Most farmers with under 1,000 trees work off-farm part-time but mostly in agriculture Labor scarcity complaints from smallholders only near Comalcalco petroleum zone; from largeholders everywhere 1975-78: failed attempt to bring in outside workers	\$20 per jornal less than in eastern Contalpa
Chontalpa Ejidos	By 1980 several wells drilled within ejidal zone; many more are planned	Labor requirements in agriculture doubled over 2-3 years Now minimum labor requirements in cacao on many ejidos; nevertheless dependent on outside workers	"Libres" paid \$75 per jornal plus housing, social security and other benefits
Teapa	Several wells drilled in area; more planned Teapa is important source of	Large number of men aged 16-36 participate in off-farm employment; severe scarcity	Similar or lower wages than in Huimanguillo Largeholders unwilling to

Teapa cont'd	labor for petroleum-related construction	for largeholders Significant new labor demand on ejidal cacao farms due to improvements and higher prices	pay competitive wages to attract workers
Coast	Major disruptions from pipeline construction and development of petroleum port	Most severe labor shortages in state	
Huimanguillo	Considerable drilling activities Rare complaints re pollution High employment of local population in nearby Cactus, Chiapas	Farmers still tend cacao, but jornaleros have gone to PEMEX, causing shortage for largeholders Large no. subsistence maize producers now work off-farm	$70-100 for 7-8 hour day

	Migration	*General Cacao Conditions*	*General Economic Conditions*
Eastern Chontalpa	Widely reported among farm families, usually temporary In northern Cunduacan most workers from farms with under 1,000 trees migrate temporarily for part-time work	Many complaints about pollution Anger over indemnification procedures and payments Much planting of new cacao	Inflation most acute here Land values have risen wildly
Río Seco	Smallholders do not seek work far away	More intensification of production, rather than new planting Most fertile area for cacao in Tabasco Long-active cacao associations Small- and medium-sized producers claim cash is greater constraint than labor	Most prosperous cacao region

Table 3.1 continued

	Migration	*General Cacao Conditions*	*General Economic Conditions*
Chontalpa Ejidos	Little emigration of ejidatarios or libres for PEMEX work Sons of ejidatarios and some libres have worked for constructoras; one estimate was 1000 per year	Highest profits for ejidatarios come from cacao; thus general desire to increase production Half of ejidatarios with rights are said to be planting cacao on private plots	General expansion of ejidal economic activity
Teapa	Much daily and weekly travel to PEMEX construction sites throughout state	Most largeholders have switched from cacao to livestock; production is dominated by ejidos Lower yields due to poorer natural conditions	Nonagricultural economic activity up; improved roads, etc. Subsistence farmers have abandoned plots in large numbers
Coast	Much temporary migration locally	Much more interplanting of cacao with coconut Problems with older plantations Big effort to organize producers for credit and processing activities	Huge increase in population and economic activity Much agricultural land has been cleared for construction projects
Huimanguillo	High temporary migration for PEMEX and constructoras Cacao workers coming in to substitute for migrants, from northern Chiapas, southern Veracruz		Most new land developed is going into livestock production

holding was much reduced. Statewide, the average holding in cacao was under three hectares. Furthermore, use of hired labor for cacao was extremely common due to growing numbers of landless workers and minifundistas.

THE CACAO BOOM OF THE 1970s

Throughout this period of structural change (since the 1940s), area in cacao and production continued to increase steadily, mainly through incorporation of new lands due to new settlement and improved communications. Charts 3.3-3.5 illustrate these trends.

During the 1970s cacao expansion took on a different character. Table 3.2 illustrates the contrast between numbers of producers, area in production, total production, price, and extent of fermentation in 1970 and 1980. Total production in 1980 was 14 percent greater than in 1970 (a fairly good year); 45 percent greater than 1968. These developments reflected the importance of changing relative prices, political organization of cacao producers, and supportive state investment programs.

A Price Revolution

Charts 3.6 and 3.7 illustrate cacao price changes in the past decades on the international and national markets. Apart from the two "lows" in world prices in 1969-61 and 1964-65--the latter offset for Tabascan producers somewhat by government guarantees--prices trended upward for producers, though not greatly in real terms. The real increases occurred in the 1970s. During that decade the national market price rose from \$6.50 to \$71 per kilo; the international price from US\$.33 to US\$1.44 per pound, with a peak in 1977 of US\$1.97 per pound.

The difference between the national and international prices was due to the system of managed prices since 1961 in the Mexican cacao industry. A yearly meeting of the National Cacao Commission (CONADECA) with representatives of the Union of Cacao Producers, industrial users of cacao and the Secretary of Commerce set the internal price of cacao and the portion of supplies to be reserved for national industry. Since 1974, these meetings have generally honored the cacaoteros' requests for prices commensurate with their costs of production, high by international standards. Industry was forbidden to import foreign cacao.

The devaluations of the peso in the mid-1970s were a great boon to the cacao export market, as the internal price was worth in U.S. dollars nearly double the former quantity of

Chart 3.3 Cacao Area Harvested in Mexico, 1925-80

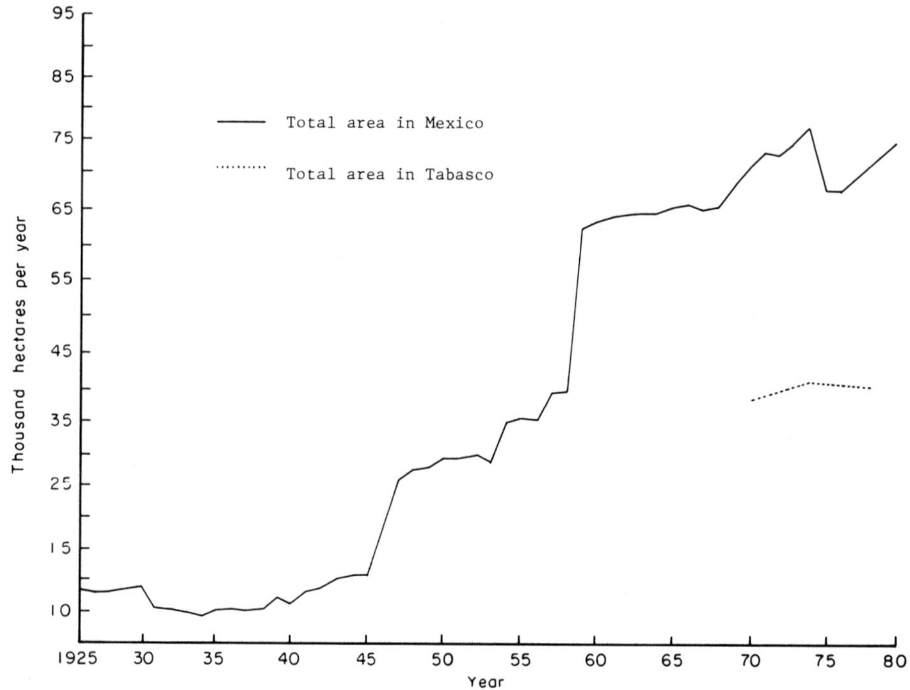

Source: Secretaría de Agricultura y Recursos Hidráulicos, Dirección General de Economía Agrícola, *Econotécnia Agrícola*, September 1977, "Cacao" and UNPC, *Boletín Informativo* No. 15, January 1983.

Chart 3.4 Cacao Production in Mexico, 1925-83

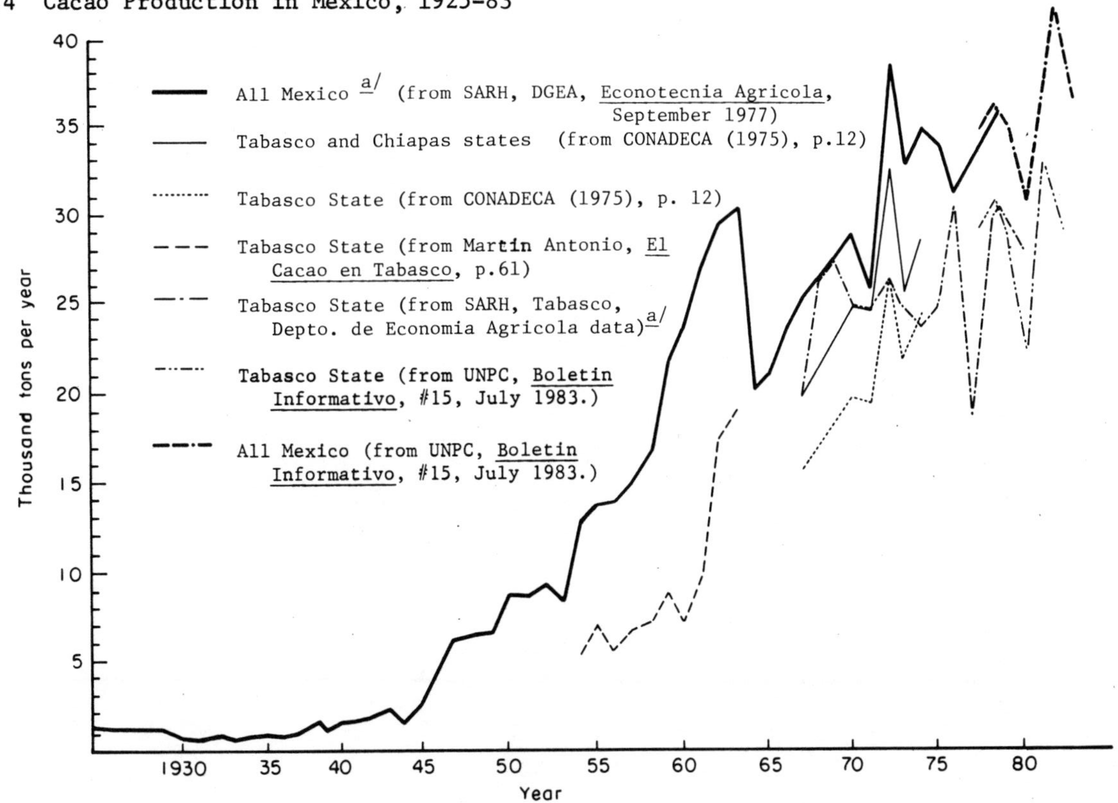

[a]Data for 1976-83 was taken from the Unión Nacional de Productores de Cacao, *Boletín Informativo* for respective years. UNPC data is provided for the agricultural, not calendar, year.

Chart 3.5 Cacao Yields in Mexico, 1925-80

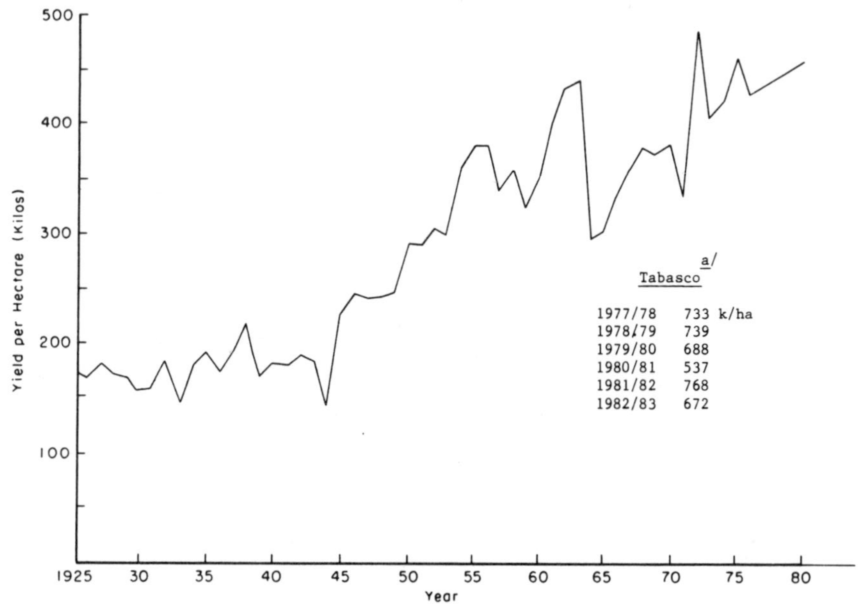

[a]Data from Unión Nacional de Productores de Cacao, *Boletín Informativo No. 15*, January 1983. Based on rough estimates of area harvested.

Source: Mexico, Secretaría de Agricultura y Recursos Hidráulicos, *Econotécnia Agrícola: Consumos Aparentes de Productos Agropecuarios para los Años de 1925 al 1980*, Vol. I, No. 9, Mexico D.F., September 1981.

Table 3.2 Changes in the Tabasco Cacao Economy, 1970-80

	1970	*1980*	*Change (%)*
Producers	11,800[a]	13,000[b]	+10.2
Area in production (hectares)	38,000[c]	42,000[c]	+10.5
Total production (tons/hectare)	19,500[d]-24,900[e]	28,465[f]	+14.3-46.0
Average cacao price, beans			
International market, NY *US$ per pound*	.30[g]	1.50[g]	+400
National market *M$ per kilo*	6.50[h]	44.00-71.00[i] (53.00-73.00 to international market)	+577-992 +715-1,023
Percent beans fermented	6[j]	51.4[j]	+757

[a]Emilio Bracho Roblín, Ing. Guillermo Vázquez Rodríguez, Ing. Marco Antonio Wing Martínez, "Situacíon del Cacao en México," Fondo de Garantía y Fomento para la Agricultura, Ganadería y Avicultura, Mexico 1972, p. 72.

[b]Comité Promotor de Desarrollo Socioeconómico del Estado de Tabasco, Plan Estatal de Desarrollo (Segunda Versión), Plan Sectorial: Agropecuario y Forestal, December 1979. Note that the UNPC *Boletín Informativo*, July 1983 reported there were 16,000 producers in the state.

[c]Unión Nacional de Productores de Cacao, *Boletín Informativo No. 10*, January 1981.

[d]Mexico, Comisión Nacional de Cacao (CONADECA), *Indicadores Económicos del Cacao*, 1975, p. 12.

[e]Secretaría de Agricultura y Recursos Hidráulicos, *Econotécnia*.

[f]Unión Nacional de Productores de Cacao, *Boletín Informativo*, September 1980.

[g]International Cocoa Organization, quoted in Unión Nacional de Productores de Cacao, *Boletín Informativo No. 4*, June 1979.

[h]Secretaría de Agricultura y Recursos Hidráulicos, *Econotécnia*, September 1977.

[i]Unión Nacional de Productores de Cacao, *Boletín Informativo No. 10*, January 1981.

[j]Mexico, Comisión Nacional de Cacao, *Indicadores Económicos del Cacao*, 1975, pp. 14, 16.

Chart 3.6 Cacao Prices in Mexico, 1925-83[a]

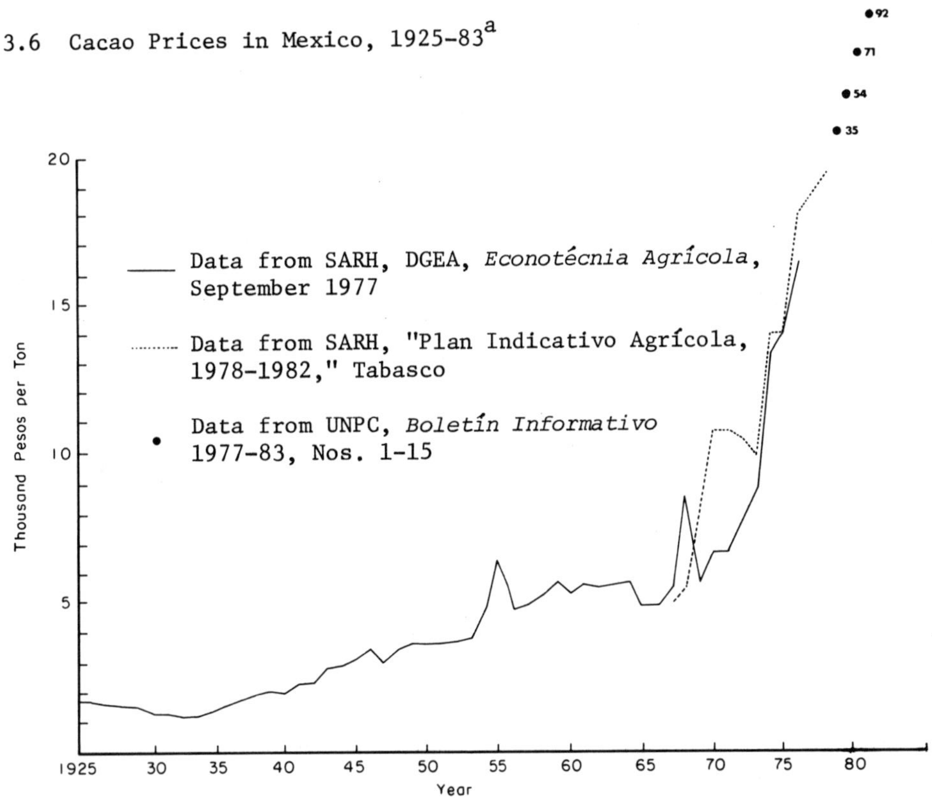

[a]Note that what appear to be very high prices 1980-82 in fact reflect declining real prices.

Chart 3.7 Average Monthly Price of Cacao, 1970-83

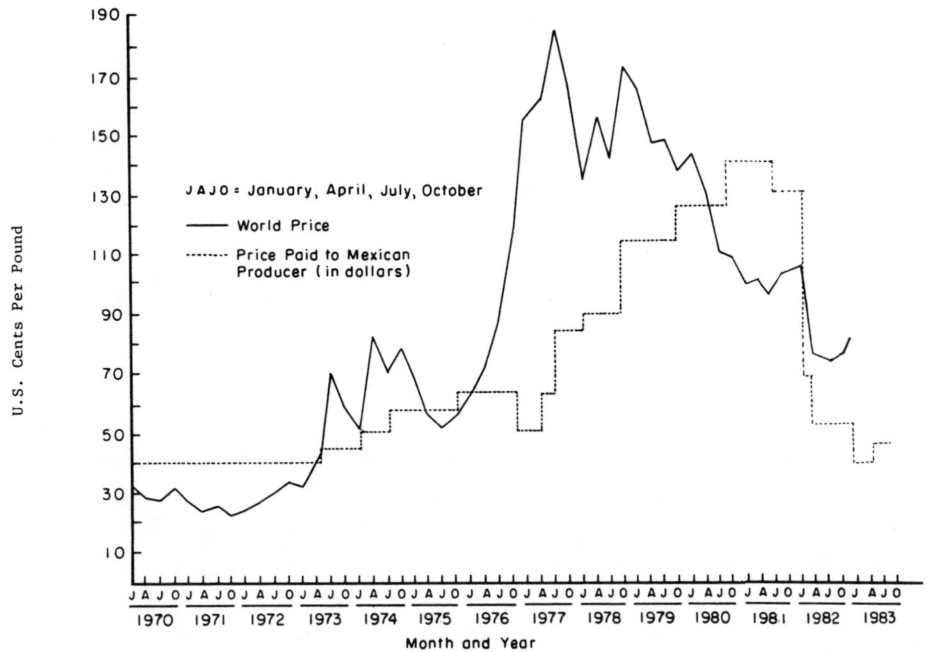

Source: Organización Internacional del Cacao, *Boletín Trimestral de Estadística del Cacao*, "Precio Promedio Mensual, Cacao en Grano," reported in UNPC, *Boletín Informativo No. 4*, June 1979, *No. 10*, January 1981, and *No. 15*, January 1983. The Mexican producer price figures after the 1981 devaluation are not exact, due to fluctuations in the dollar value of the peso.

pesos. Producers chafed at the large part of the cacao supply promised to national consumers at a price well below the international price. In 1979-81, however, farmers received well over the plummetting world price from their industry consumers, and most production was sold nationally. The large devaluation of the peso in February 1982 was greeted with satisfaction by the industry as a whole, which claimed that during the prior period exports had been constrained because local costs were higher than the sales price in national pesos (UNPC, Boletín, March 1983).

Price increases for cacao during the 1970s were so dramatic as to make fundamental changes in the standards of living of cacao farmers. Their material culture became "modernized." Children were taken out of the regular work force to go to school and had achieved education and job opportunities unthinkable for their older brothers and sisters.

Cacao Producer Organization

The first cacao producers' organization was set up in the 1930s, as part of Governor Garrido's farmer mobilization, primarily functioning as a credit union. Early attempts to improve cacao quality and processing and control planting in anticipation of a world glut (recommended by the "Trinidad Commission") had little impact in the face of Río Seco pioneer settlement and extensive planting of low-quality, high-producing varieties. The producer organization broke up in 1953 when cacao prices and production levels increased and the need for joint action seemed less urgent.

Achievements of the 1960s. But the anticipated glut indeed began in 1960. In response, the government of Mexico set a support price for the first time--of $8.60 per kilo for the internal market--and set up import barriers to protect national production (Bracho R. 1972, p. 20). In December 1961, with Governor Madrazo's encouragement, a producer's union with more far-reaching powers was devised to relieve the chaotic results of the 1960 price tumble and to take advantage of future international export potential.

The National Union of Cacao Producers (UNPC) began to operate in 1963 with 20 local Tabasco associations and 17 from Chiapas. An individual joined a particular association because of geographic proximity, credit availability, and trust in the leadership; but he could deliver cacao to any association.

Funding from UNPC associates was used to build the Industrializadora de Cacao de Tabasco, S.A. (INCATABSA), the first modern Tabascan chocolate factory, in 1960-64, with a capacity of 10,000 tons of dry cacao per year (Antonio

Ortíz 1964, pp. 68-71). Prior to this time, there were only three, much smaller chocolate processors in the state (Bracho R. 1972, p. 59).

In 1962, Kennedy's Alliance for Progress program funded FIRA in Mexico to increase credit for cacao production. The government of Mexico and UNPC began a new program of research and extension in 1964 to boost cacao productivity. Grafts and "clonal" cacao were popularized; new shade trees recommended; and a program was set up to encourage use of chemical pesticides. Fertilization was introduced but was as yet uneconomical and unaccepted (Antonio Ortíz 1964, pp. 23-25).

In 1968, FIRA-backed private bank loans were made for the Río Seco Association to build the state's first large processing plant for postharvest fermentation and to improve bean quality (Bracho R. 1972, p. 94). Between 1964 and 1969 total cacao exports--mainly to the United States--rose sevenfold, though nearly all Mexican cacao was primitively processed (only washed and sun-dried).

The 1970s: Market Diversification and Vertical Integration. In its first decade UNPC marketed cacao mainly through private financiers. But as a result of the drop in prices in 1968-69, the newly independent Union found itself near bankruptcy and unable even to pay the advance to farmers upon delivery. In response the government gave UNPC direct control over $80 million pesos worth of cacao held by comisionistas, and marketing credit worth $95 million was arranged with the Banco Agropecuario del Sureste, S.A. ("Alianza Popular" 1974, pp. 9-12).

Thereafter, UNPC directly marketed cacao and INCATABSA-produced cacao subproducts to industrial and international consumers. When cacao was brought to the association (or fermenting plant), it was weighed and the farmer was immediately paid an "anticipo" somewhere near the guaranteed price with borrowed funds. The "remanente" was paid after the cacao was sold, often months later. It was the remainder left over after the anticipo, marketing costs, and Union administration expenses were deducted from the sales price. The remanente was basically dependent upon the level of export sales. Chart 3.8 shows the levels of anticipo and remanente between 1973 and 1980.

As a result of aggressive marketing, Mexico exported to 18 countries by 1977. At the end of 1979, Mexico was the seventh most important cacao producer in the world (UNPC, Boletín No. 6). CONADECA and UNPC hoped eventually to limit exports entirely to processed cacao products, rather than beans, and to ferment all beans sold internally. This policy seemed wise: within a few years, new plantings in Brazil and Ivory Coast were expected to come into production, and this was likely to coincide with a prolonged period of weak demand due to depressed world economic conditions.

Chart 3.8 Sales Price on the National Market and "Anticipo Libre" to Producers for Washed Cacao 1973-80

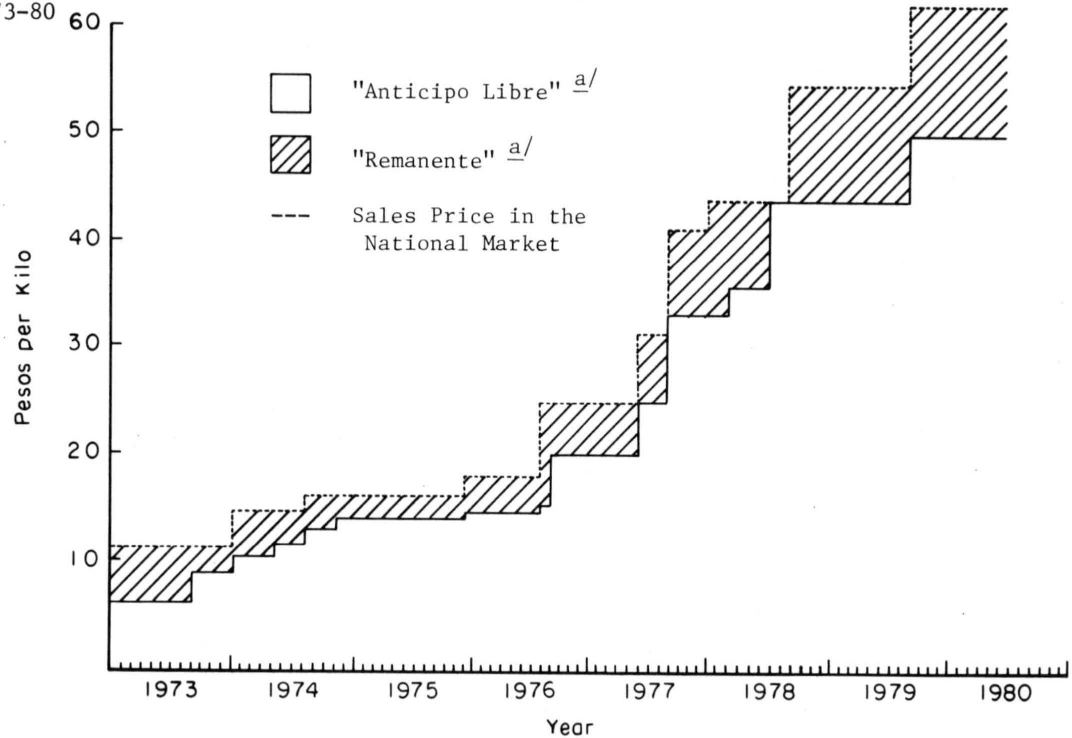

[a] See text for definitions of "anticipo libre" and "remanente."

Source: Mexico, Unión Nacional de Productores de Cacao, *Boletín Informativo No. 1*, November 1978, p. 13 and December 1980; CONADECA, "Indicadores Económicos," 1977, pp. 14-15.

In 1980, with a highly overvalued peso, 57.1 percent of Tabasco cacao was sold to the national market; .8 percent for export of beans; and 29.7 percent was sent to INCATABSA for cocoa butter exports and cocoa powder sales to the national market. In 1981-82, 83.4 percent went to national consumption; a year later, after the devaluation, only 38.5 percent was purchased domestically. Unfortunately, November 1982 saw the lowest monthly price since March 1976 in the international market (UNPC, Boletín July 1983).

Between 1975 and 1977, eight fermenting centers were built and others begun. By January 1981, fermentation capacity had been expanded to 30,000 tons of dry cacao beans, almost enough to meet farmer demand at peak harvest periods. In 1975, only 20 percent of cacao was fermented; in 1980, 51 percent (UNPC, January 1981). In 1981, a new program of "minifermenters" was developed, allowing small groups of producers with a combined annual production of 200 tons of dry cacao to form a credit society to build on-farm facilities (UNPC, August 1981).

In 1978, a cacao drying facility with capacity for 27,850 tons was built. By 1981, INCATABSA capacity had expanded from 6,500 to 12,500 tons and was operating at 72 percent capacity. New distribution, storage and transport facilities, and new headquarters in Mexico City were built. Construction of a new plant to process cacao subproducts was proposed. The UNPC also joined national industries in a program to encourage domestic chocolate consumption.

A "Program for Economic Assistance to the Cacao Sector" was developed by UNPC (assisted by IMPECSA and CONASUPO) in the early 1980s to provide cooperatively purchased basic foods, tools, and inputs to keep producer costs down. The pilot project with three associations was considered successful, and the program was to be expanded.

Federal and State Support for the Cacao Plan

The state and federal governments supported the cacao producers of Tabasco by providing tax relief, construction of needed infrastructure, support in international marketing, support in struggles with PEMEX, and cooperation with the UNPC for a major technical assistance program, the Cacao Plan.

In November 1980, as part of SAM, state and municipal taxes on agricultural production were lifted. The anticipo rose from $53 to $57.17 per kilo. Also in 1980, the national government returned all indirect and general export taxes paid for cacao exports. Mexico was a party to the Third International Cocoa Agreement and lobbied unsuccessfully for its acceptance by international consumers.

The Tabasco State Plan program and federal investments in the state yielded a number of new infrastructure projects useful to the cacao economy. Roads were built, improving both farm-to-market transport and national marketing. Electrification and education investments (particularly in primary school and agronomy studies) assisted cacao farm families. The Cacao Plan gave concrete form to the political backing given by the federal and state governments to the cacaoteros. It was not a direct response to the "boom," but part of a nation-wide program inspired much earlier in the decade to plant cacao throughout southern Mexico, increasing total area by 50-100 percent (personal communication, Turner Price, September 1982). It demonstrated serious support for agriculture at the very time when petroleum development was most threatening to the agricultural sector.

Table 3.3 compares the technical characteristics of Tabascan cacao farms in 1978 with the technical recommendations of the SARH. The chief problems appeared to be insufficient weeding, fertilization, and pruning; a poor age distribution of cacao trees; and poorer than average condition of cacao and shade trees. The Tabasco program was developed to establish new plantations; rehabilitate older plantations; provide technical assistance to improve yields and farmer incomes (mainly by promoting increased fertilizer use, better management, and revised insecticide practices); and carry out research. Its specific goals were to increase yields per hectare from .641 tons (average 1974-80) to 1.3 tons by 1982 on rehabilitated plantations; to rehabilitate 34,900 hectares 1979-82 (i.e., three quarters of all Tabascan cacao); to establish 2,000 hectares per year of new plantations--10,000 hectares in all by 1982; and to achieve production of 49,800 tons in 1982 (Plan, p. 17).*

Eighteen nurseries were set up throughout the state providing stock free of charge; by 1981 1,850 producers had received cacao and shade plants. Between 1977 and 1980 6,000 hectares of cacao had been planted and 15,000 hectares had been rehabilitated.

The technical assistance program was set up under a SARH Rainfed Agriculture District. In 1979 seven agronomists and

---

*Planners estimated that without the Plan, production in 1985 would reach only 32,558 tons, worth $1,771.2 million; with the Plan they hoped for 57,100 tons worth $3,106.2 million, or a total increase of 75 percent in production, value, and farmer incomes. Without the Plan, Production would reach a plateau of 33,358 tons in 1982; with the Plan, 63,700 tons in 1990 (SARH 1980, pp. 63, 67). Over a ten-year period (1979-88), total returns would be 56 percent higher with the Plan. The program would cost an extra $4,268 million, but it would increase the value of production by $9,601.3 million, not including multiplier effects to local economies.

Table 3.3 Characteristics of Tabasco Cacao Production, 1978 *(in percent unless otherwise specified)*

Characteristics	1978[a]	Recommended[b]
Average number of trees per hectare	567.74	500-600
Average production per hectare *(kilograms)*	538.50	1,000-1,300
Area with credit	6.01	20-25
Ejidatario producers	37.30	--
Private property producers	62.70	--
Pruned trees	84.98	100
Producers who fertilize	46.40	100
Producers with spraying equipment	64.00	100
Who use: manual sprayer	56.90	
motorized sprayer	42.90	
Use of insecticides:	80.80	100
Metilico	63.27	
Lindane	13.70	
BHC	12.50	
Parathion	8.90	
Use of fungicides:	80.80	100
Copper sulfate	94.00	
Tribasic sulfate	2.70	
Cupravit	1.90	
Locide	1.10	

Table 3.3 continued

Characteristics	1978[c]	Recommended[b]
Condition of shade trees:		
Good	37.40	100
Regular	55.70	
Bad	6.80	
Farm needs drainage	26.65	
Farm has drainage (of those that need it)	63.80	100
Condition of cacao trees:		
Good	43.10	100
Regular	51.20	
Bad	5.60	
Farmer has cacao nurseries	5.80	Not necessary since UNPC has set up nurseries
Farm access to all-weather roads	54.70	100
Farmer is literate	82.30	100
Types of cacao planted:		
Criollo	67.40	Highest quality but low yields
Calabacilla	80.30	
Amelonado	69.20	
Clonal	82.40	
Guayaquil	78.80	
Types of shade trees planted:		
Moté (Erythrina)	76.20	Hope to cut down, due to pest infestation

Samán	1.70	
Cocoite	12.10	Want to increase plantings
Chipilcoite	11.10	
Age distribution of trees:		
0-5 years	8.60	Completely replace trees after 30 years; have continuous cycle of new trees coming into production
6-10 years	34.90	
11-15 years	25.30	
16-20 years	30.60	
21-25 years	0.30	
26-30 years	0.20	
31-40 years	0.03	
50 years plus	0.03	
Yearly mortality of trees in new plantations		Two to five percent
Weeding:		
Once a year	21.80	Three or four times a year or enough to keep clean
Twice a year	49.00	
Three times a year	19.00	
Four times a year	10.00	
Pruning[b]:		
2-3 small branch prunings a year	60.00	100
One large pruning a year	51.00	100 every year or two

[a]SARH, "Censo de Cacao," Villahermosa, Tabasco, 1978.

[b]SARH, "Plan de Incremento de la Producción de Cacao en el Estado de Tabasco," Villahermosa, Tabasco, January 1979.

50 agricultural technicians were working with the cacao farmers specifically--sevenfold more than before the Plan. Still, the average area covered by an assistant was 500-600 hectares. As of January 1981 some type of extension contact had been made with 10,800 cacaoteros on 30,239 hectares. In addition to the programmed four to five yearly visits to each farmer involved, a weekly radio program was aired by the head of the Plan which was listened to enthusiastically by every cacaotero surveyed. By 1983, 50 demonstration plots were planted in farmers' fields, and research was being carried out on experiment stations.

To achieve these goals, it was estimated that 81 percent of costs would be borne by the producers; the rest by SARH. Table 3.4 compares Cacao Plan figures for the costs of production of cacao cultivation with and without rehabilitation. These budgets did not include costs for interest on loans, depreciation for farm equipment or buildings, cacao transport, or on-farm processing. Opportunity costs for farmers' labor were assumed to be the same as an agricultural worker's wage. Land was grossly undervalued at $4,000 per hectare. With "technification," labor use was estimated to rise by 71 percent, current costs by 75 percent; yet the difference in profits (even by these optimistic budgets) was only 5.4 percent, despite a 30 percent estimated yield increase.

Given these statistics it would seem rather remarkable that large numbers of farmers indeed adopted parts of the package. In fact, those producers who were the principal adopters of the new technologies, faced a far different real budget structure. Those producers whose budgets *did* more or less correspond to the Cacao Plan figures did *not* adopt the technologies because the profits were not competitive with other economic options.

The Cacao Plan as a whole was in 1982 still far from reaching its ambitious goals, although there had been substantial rehabilitation and new tree plantings. Given the economic conditions in Tabasco, these were significant accomplishments; without the petroleum boom, one would have expected an even more dramatic increase in state cacao production.

IMPACT OF THE PETROLEUM BOOM ON THE CACAO SECTOR

The economic disruptions from PEMEX activities by no means bypassed the cacao farmers. Inflation and damage from pollution affected all farmers negatively, while labor scarcity, production changes, and political unrest varied significantly in their impact according to family characteristics.

Table 3.4 Costs of Production for Cacao Cultivation in Tabasco State

Activities	*Without Rehabilitation* Cost	*Without Rehabilitation* Man-days	*With Rehabilitation* Cost	*With Rehabilitation* Man-days
	(pesos)		(pesos)	
Weeding	996.00	12.0	1,992.00	24.0
Pruning	747.00	9.0	747.00	9.0
Fertilization	207.50	2.5	830.00	10.0
Pest Control	996.00	12.0	1,660.00	20.0
Disease Control	1,245.00	15.0	2,490.00	30.0
Drain Maintenance	249.00	3.0	498.00	6.0
Purchase of Pesticides	1,047	---	2,365.00	---
Purchase of Fertilizer	1,500	---	2,366.25	---
Harvest and Pod-breaking	2,324.00	28.0	3,320.00	40.0
TOTAL	9,311.50	81.5	16,268.25	139

Yield: 637 kg./ha. 825 kg./ha.

Cost Difference: \$6,956.75

Yield Difference: 188 kg./ha. @ \$44/kg. = \$8,272

Profit Difference: \$18,715.50 vs. \$19,731.75 = \$1,015.25

Source: SARH, "Plan de Incremento de la Producción de Cacao en el Estado de Tabasco," Villahermosa, Tabasco, January 1979, p. 71.

Direct PEMEX-cacaotero confrontations took place over the company's practices of expropriation, trespass, and ecological destruction. Polluted waste waters from oil drilling ran off onto cacao plantations, and construction of pipelines and roads often plugged essential agricultural drainage canals. There was a real increase in levels of personal violence--within families, in communities, between strangers --as a result of increased tensions, frustration, and insecurity.

Yet the level of political violence among cacaoteros was not marked. One would have expected there to be far more. The political system involving Tabascan cacaoteros seems to have worked to control conflict.

Most of the "uprisings" reported were, in fact, concerted political moves, by politicians well in control, to show strength in the face of PEMEX or federal intransigence on some issue. Though violence was at times threatened on one side or the other, there were few examples of either the government's unleashing the national guard or police with deadly weapons, or violent destruction by campesinos of PEMEX property, or threats to personnel. One cannot help but wonder if the same would have been true had petroleum development taken place in central Chiapas or Oaxaca, which have a history of campesino repression.

How did the farmers themselves perceive these conflicts? There was little ambiguity about the anger of farmers whose lands had been damaged, who still awaited compensation. Many farmers in drilling areas were truly concerned about the long-term ecological damage of PEMEX activity and tended to blame pollution for problems on their plantations. Farmers near the pipelines and Dos Bocas port were concerned about the effects of a potential oil spill.

But for most--even those with relatives working on PEMEX projects--the oil company was very remote and its impacts nebulous: the poorly understood cause of inflation (prior to the boom, most consumer prices tended to fluctuate with the cacao price); and the source and object of congested highways on the way to Villahermosa. Anger was found mainly among populations in the Reforma area and parts of Cunduacán and Paraíso directly involved in intensive drilling or processing.

Even this anger was relatively controlled, as much because there were effective outlets for it as because of existing political repression. The UNPC associations were forums immediately available for airing and sharing complaints, and the leadership did put its grievance system into action. Municipal, ejidal, and ranchería leadership also supported farmers with the governor and others. Decisions to bar PEMEX trucks and other direct actions seem to have been well-organized community decisions approved by some cooperative, political, or community group when other routes of action were ineffective.

Labor Scarcity and Migration

The effectiveness of cacaotero political action, like that of the rising cacao prices, was essential in protecting cacao producers during the oil boom. But the key linkage between cacao production and petroleum development was through the labor market, and the sector's most important source of stability lay in the structure of cacao production and labor utilization.

The supply of wage labor for cacao production became increasingly scarce, and its price increased as a result of increasing general wages, increasing cacao prices, and worker migration to petroleum zones. Before the petroleum boom, labor prices rose more or less parallel to the cacao bean price, at a rough ratio of one kilo:one jornal (a day's work). As young men flocked to jobs in petroleum and construction, the price of labor was bid upwards. By 1980 the cacao:wage ratio had risen to three kilos of cacao:one jornal.

The Sociodemographic Survey documented a significant exodus away from cacao work. Between 1965 and 1979, 20 percent of the EAP with some history of cacao activity (accounting for 69 percent of the total EAP surveyed) had ceased working in cacao. Five percent of all who had ever owned cacao land and 32 percent of those who had worked in cacao, but never owned land, left the sector. (These figures included a small number who had simply retired, as well as those who changed occupations.)

A fifth of the surveyed population over seven years of age had experienced a change of residence at some time in their lives. Between 1970 and 1980, 68 percent of all migrations took place. Surprisingly, however, if we break up this figure into the periods before the boom (1970-75) and the period after the boom (1976-80), we will find that significantly more moved in the earlier period before cacao prices rose. Migration of men over forty particularly declined in the latter part of the decade.

Cacao Production

These changes in the price and distribution of labor, the most important input in cacao production, markedly affected production possibilities for different types of producers. Landless workers benefitted from the labor situation, as did minifundistas who depended heavily for cash on off-farm employment, although many of them chose to work in petroleum rather than agricultural off-farm jobs.

This situation certainly challenged the Cacao Plan goal to raise labor inputs in production by over 70 percent for the package of improved technology. Basically, only those farmers who could minimize cash costs--by utilizing family labor in place of hired labor and substituting family labor for cash inputs like pesticides--found it economically rational to adopt the program.

Thus cacao production from family-labor-based farms did increase as family members' employment off-farm decreased because of better returns from on-farm work. Minifundistas maintained fairly stable production in most cases, quite limited in their ability to purchase cash inputs, and they frequently switched from jornalero to nonagricultural off-farm employment to meet cost-of-living increases.

Farmers dependent on hired labor reduced production. Their former minifundista and landless workers preferred work for PEMEX; and family-labor-based farmers preferred to intensify family cacao production. They were further induced to reduce their efforts in cacao by the sudden availability of numerous investment alternatives created by the petroleum boom.

FAMILY ECONOMY PROFILES OF TABASCAN CACAO FARMERS

For purposes of analyzing the impact of petroleum development on the cacao economy of Tabasco, I chose to disaggregate the cacao farm households on the basis of their relation with the labor market, as outlined above. Both personal field research and analysis of the Tabasco Cacao Census and Sociodemographic Survey confirmed that the labor:land ratio was the most important determining variable of production response. The groups were defined as follows:

Minifundia had under two hectares of cacao (1,000 trees) --the amount a single person with occasional family help could manage. This group accounted for about half of producers and in 1978, 30 percent of the cacao produced in the state. Off-farm employment was extremely important for this group. Minifundia predominated in densely-populated Comalcalco, Jalpa and Paráiso municipios and were largely (85 percent) propietarios.

Family-labor-based producers with two to six hectares of cacao (1,000-3,000 trees) were generally able to cover labor demands with family workers. Labor was hired only during especially busy harvest periods, at times when some family members worked off-farm and had to be replaced in a crisis, or when the family was at a life-cycle stage with few working members. Over a quarter of Tabascan cacaoteros were family-based; they produced about a third of the state's cacao in 1978.

Those with over six hectares of cacao depended upon hired labor. These *hired-labor-based* producers had a wide range of economic alternatives to cacao production because of their greater resource base. About ten percent of Tabascan cacao farmers were dependent on hired labor, but in 1978 they had accounted for nearly 30 percent of production.

A fourth type of producer belonged to *collective ejidos* and worked in both collective plantations and cacao on personal plots. The collective used separate criteria to hire ejidatarios, with and without land rights, and the landless; in general, the collective had many alternatives to cacao production. The personal plot often operated along the same lines as minifundia production.* In fact, the vast majority of collective production came from the collective plots. This group numbered approximately 15 percent of total producer numbers (as individuals), but as some were only marginally involved in cacao, this overstated their importance. Collective cacao accounted for nearly ten percent of the state total.

The fifth group in the cacao economy structure was composed of *landless workers* who resided in cacao-growing communities and worked as jornaleros throughout the year on family farms and hired-labor-based plots. They probably numbered 4,000-5,000. They accounted for a fifth of all workers in cacao, compared with about 12,000 cacao farmers (52 percent of all workers), and another 6,000-7,000 farm family members who worked in cacao production.

Chart 3.9 shows the relative importance of these types of farmers in terms of numbers, area planted to cacao, and total cacao production in 1978. Analysis of the Mexican Population Census suggests that between 1970 and 1980, the numbers of minifundistas increased, and cacao economy information suggests that the proportion of cacao coming from largeholders decreased throughout the decade.

These five kinds of households--minifundia, family-labor-based (FLB), hired-labor-based (HLB), collective farms, and nonfarm families--operated under fundamentally different modes of production, conditioned by the asset array available to them and the organization of labor. The charts and graphs in this section summarize the differences, which later chapters will explain. The principal aspects reviewed here are: farm assets and means of production; farm family characteristics and the demographic cycle; family employment and relations of production; farm processes, production systems, and costs; and family strategies for labor use and farm production. Full statistical summaries of household demographic, employment, and other information from the SST

---

*Collective producers were included among the "minifundia" in the Cacao Census analysis because only their private plot production was included in the study.

Chart 3.9 Relative Importance of Tabasco Cacao Farm Types 1978-80

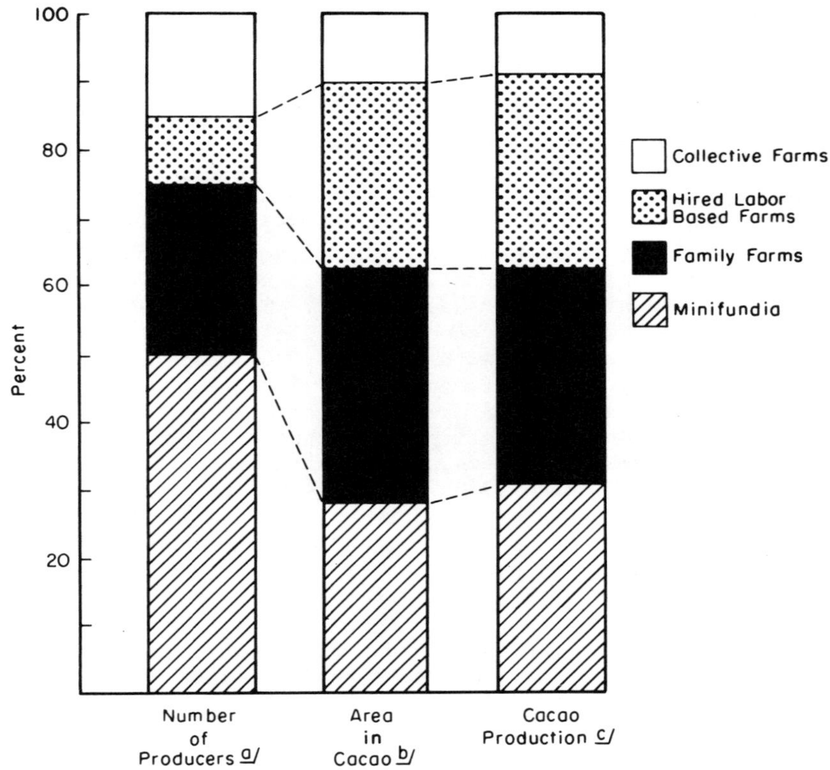

Source: Estimated from data in SARH, *Censo de Cacao de Tabasco*, 1978; COPRODET and CONAPO, "Encuesta Sociodemografica del Estado de Tabasco," Villahermosa, Tabasco, 1980; SARH, Promotora de la Chontalpa, "Cacao," 1980.

may be found in Scherr (1983b), Appendix Tables C-1 through C-23.

Farm Assets and Means of Production

The case studies and regional survey showed clear differences in the range of farm assets available to the various groups of families. It is important to contrast the value of assets in land--which were highly inflated during the 1970s--with the assets in buildings, equipment, and livestock, items far more indicative of wealth. Land belonging to minifundistas and ejidatarios on collectives ranged in value below $500,000 or US$22,000 (in highly overvalued pesos); FLB farms ranged up to US$50,000; HLB farms between $50,000 on modest establishments and US$1,000,000 on large-holdings planted heavily to cacao. The value of buildings, equipment, and livestock totalled well under US$5,000 for minifundistas, under US$10,000 for family-labor-based farmers, and from $15,000 to half a million dollars for hired-labor-based producers.

The minifundia averaged a little over a hectare in size and produced roughly a fourth of the average production on family-based farms, which averaged about six hectares. The hired-labor-based farms, by comparison, averaged about fourteen hectares, and the average farm produced between two and three times as much as the family-labor-based farm. Production per ejidatario on the collective was in the range of the family-labor-based producer.

Farm Family Characteristics and the Demographic Cycle

Cacao-producing families in Tabasco had very similar overall characteristics to the "typical" Tabascan family. Half of all family members were fifteen years of age or younger; slightly over a quarter were aged 16 to 30. Of all family members over seven years of age, 36 percent were head of household or spouse; 50 percent were children of the head of household; and 14 percent had some other relationship, commonly grandchild or daughter-in-law of the head of household.

Charts 3.10 and 3.11 illustrate the typical demographic stage for the different family types. The hired-labor-based producers were generally older, and their children tended to be older as well. The minifundistas tended to be younger and had more younger children. The population on collectives and without land was much younger than that of farm households. Median age of head of household was only 35 in non-farm families, 40 on collectives, 47 on minifundia, 49 on

Chart 3.10 Population in the Sociodemographic Survey of Tabasco Cacao Subsample, by Type

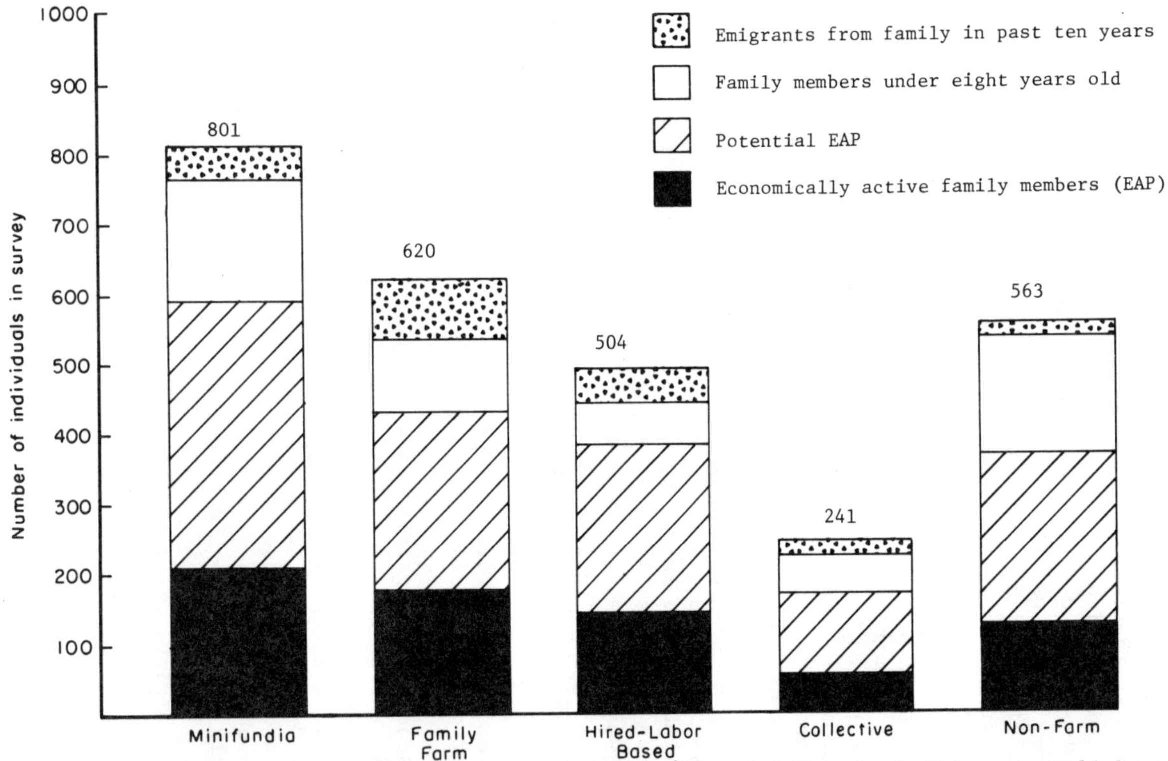

Source: COPRODET and CONAPO, Encuesta Sociodemográfica del Estado de Tabasco, Villahermosa, Tabasco, January 1981, Cacao Subsample.

Chart 3.11 Demographic Stage of Cacao Households, by Type[a]

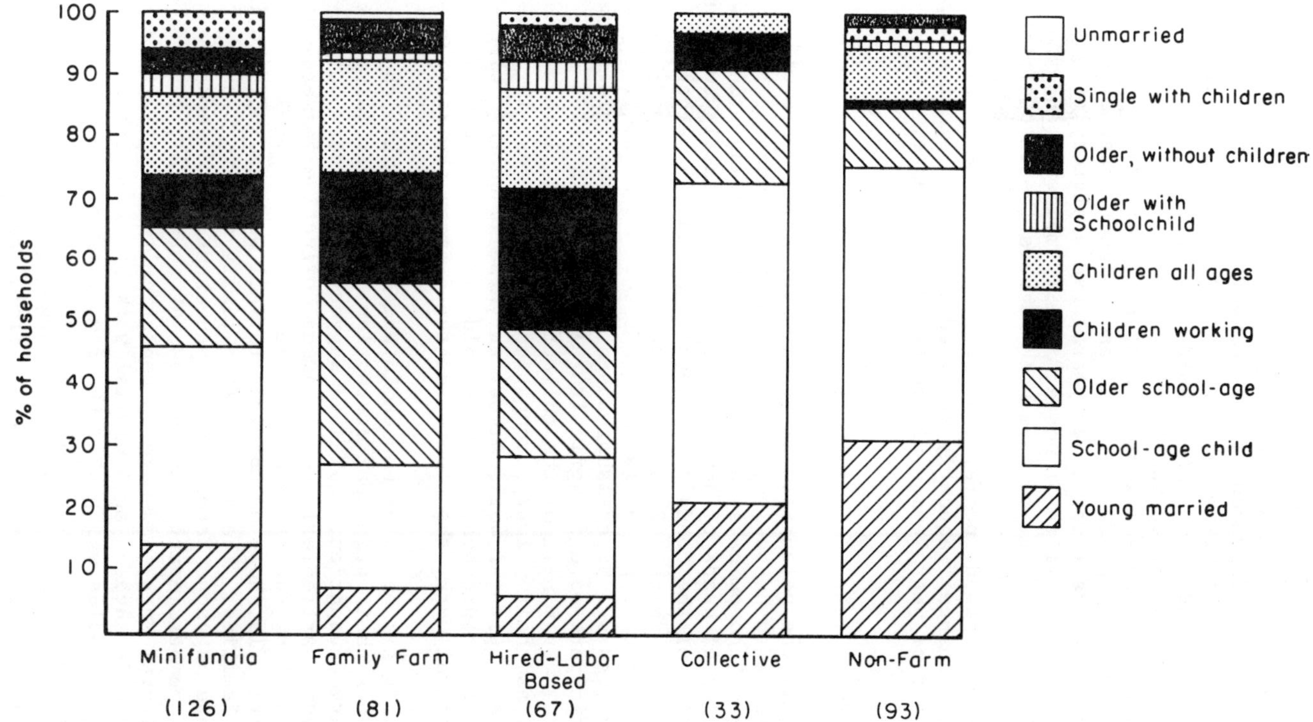

[a]Definitions for the various stages may be found in Appendix Table C-6, Scherr (1983b).

Source: COPRODET and CONAPO, "Encuesta Sociodemografica del Estado de Tabasco," Villahermosa, Tabasco, January 1980, Cacao Subsample.

family-labor-based farms and 55 on hired-labor-based farms. Largely as a consequence of this configuration, there were relatively more dependent family members in the minifundia, collective, and nonfarm families, and more family emigrants from family- and hired-labor-based households, since these were generally married children.

Chart 3.12 shows migration histories for the family types. Nearly a quarter of all family-labor-based farm household members (over seven years) had a history of changing residence, a fifth of nonfarm family members, and nearly two-thirds of collective household members (because of the dislocation caused by collectivization). Note that on family-labor-based, nonfarm, and hired-labor-based farms residence change frequently occurred after 1976, indicating the impact of petroleum employment-induced change.

Family Employment

Of all the potential EAP (those over seven years old), a quarter were housewives and a third were students; the rest were "economically active." Charts 3.13 and 3.14 show occupation of heads of household by type. A fairly constant proportion of the heads of household (around 75 percent) were cacao farmers. Relatively few minifundistas participated in other types of farming, but more were involved in skilled trades. Hired-labor-based heads of household tended to participate more in agricultural work and in trade and commerce. Only half of the collective farmers claimed cacao farming as their primary occupation, with the rest in farming or agricultural work, almost universally on the collective. Of nonfarm heads of household, 72 percent were in agricultural work, the others scattered among the other types of work.

Almost all heads of household who did not claim cacao farming as their principal occupation, claimed it was their secondary occupation. SST data suggested secondary occupations were particularly common among collective farmers, and uncommon among nonfarm heads of household.

Chart 3.15 shows occupations for all working individuals in the families. The minifundista family members tended to work more as cacao farmers (partly because there were few working members apart from the head of household), less in agricultural work, and more in nonagricultural activities. The preponderance of agricultural work by hired-labor-based farmer households was partly the result of more older children in the EAP, many of whom worked on their family's farm.

But disaggregating by age of worker in Chart 3.16 revealed very different patterns within and between the families. Minifundista workers aged 16-20 were much less

Chart 3.12 Residence Changes in Cacao Families, by Type

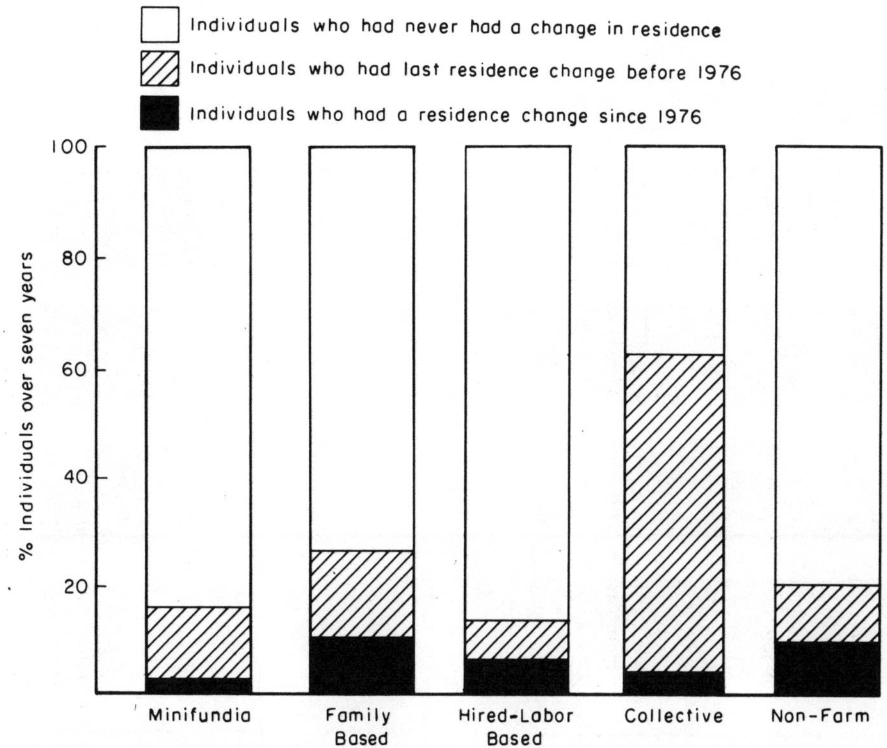

Source: COPRODET and CONAPO, "Encuesta Sociodemográfica del Estado de Tabasco," Villahermosa, Tabasco, January 1980, Cacao Subsample.

Chart 3.13 Primary and Secondary Occupation of Head of Household on Tabasco Cacao Farms, by Type

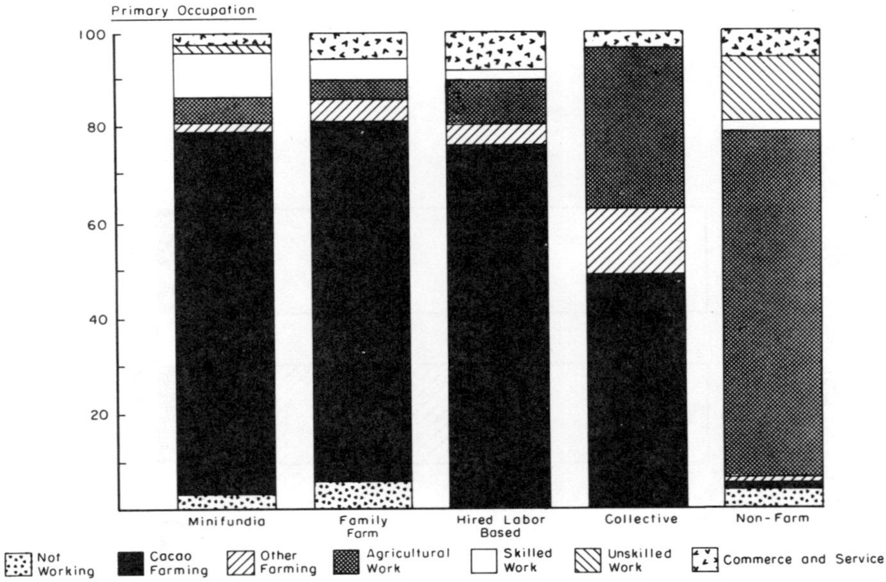

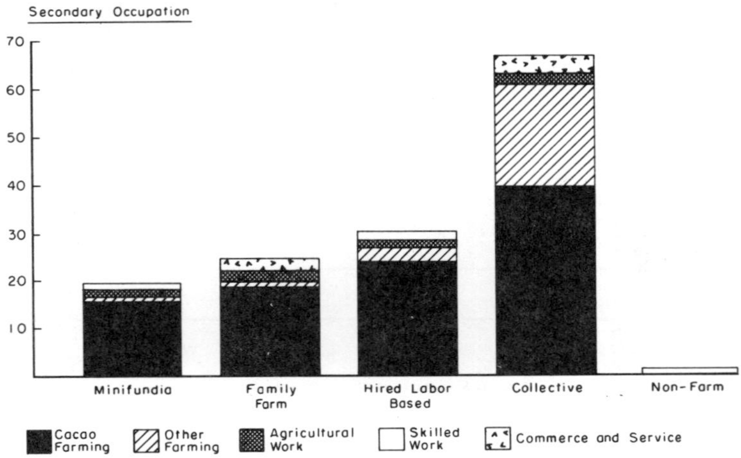

Source: COPRODET and CONAPO, "Encuesta Sociodemográfica del Estado de Tabasco," Villahermosa, Tabasco, January 1980, Cacao Subsample.

Chart 3.14 Primary Occupation of Cacao Household EAP, by Type

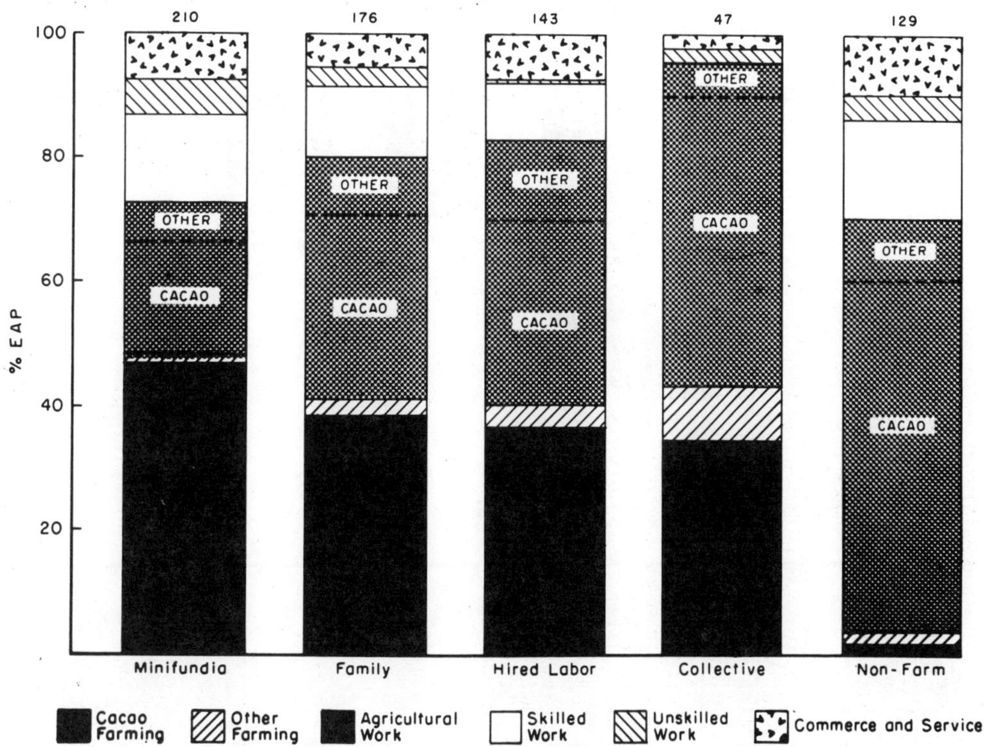

Source: COPRODET and CONAPO, "Encuesta Sociodemográfica del Estado de Tabasco," Villahermosa, Tabasco, January 1980, Cacao Subsample.

Chart 3.15 Secondary Occupation of Cacao Household EAP, by Type

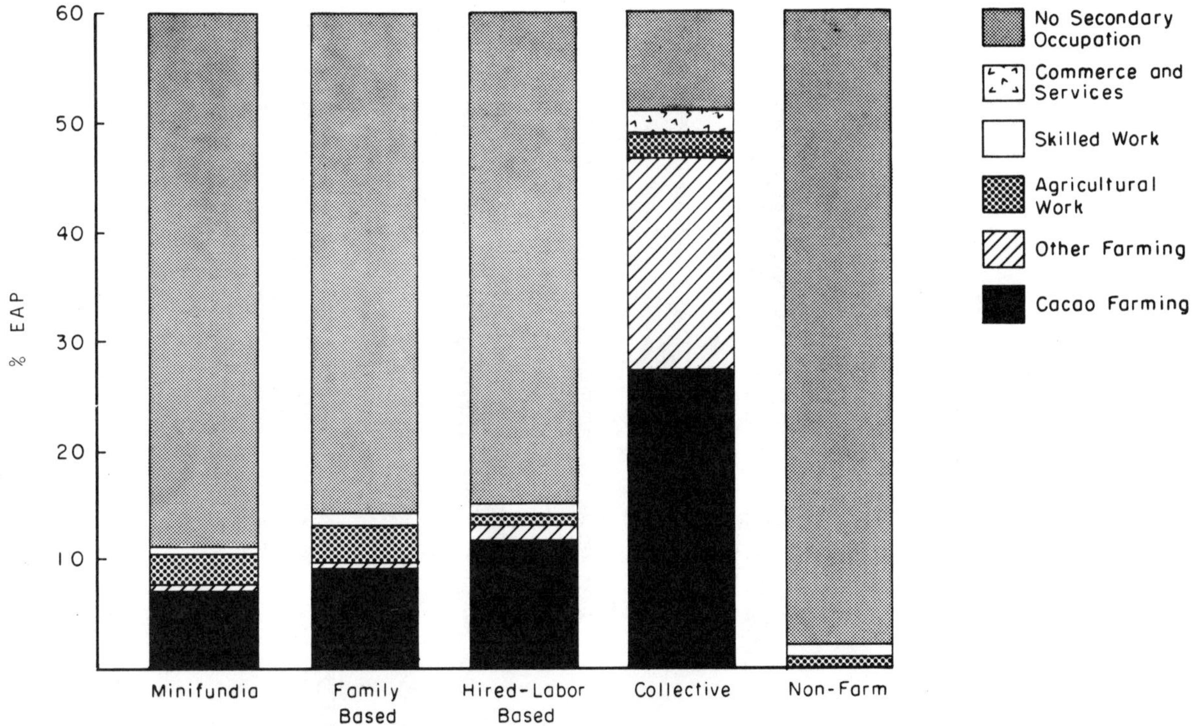

Source: COPRODET and CONAPO, "Encuesta Sociodemográfica del Estado de Tabasco," Villahermosa, Tabasco, January 1980, Cacao Subsample.

Chart 3.16 Primary Occupation of Potential EAP, by Age[a,b,c]

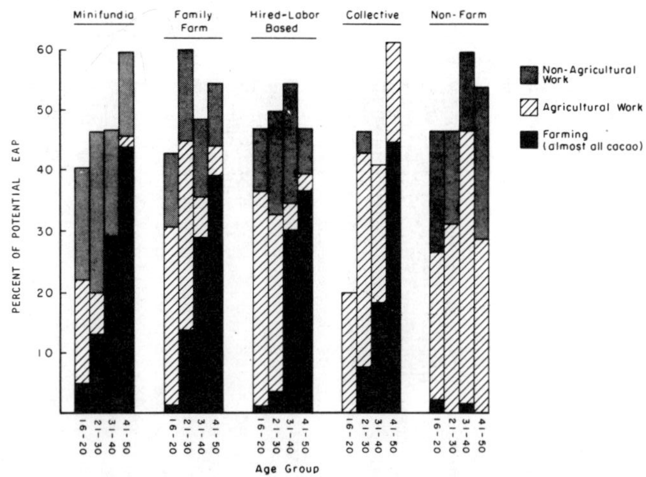

[a]Note that of 646 individuals aged 8-15, only 57 were working, all but three of whom were agricultural workers.

[b]Of 234 individuals aged 61-99, 157 were still working; all but ten of those were working in agriculture.

[c]The graphs show that a portion of the potential EAP, which was in fact employed; those remaining include include housewives, students, the sick, and retired individuals.

Source: COPRODET and CONAPO, "Encuesta Sociodemográfica del Estado de Tabasco," Villahermosa, Tabasco, January 1980, Cacao Subsample.

likely to work in agriculture than those on family or hired-labor-based farms because they were not necessary to on-farm production. Like teenagers from nonfarm households, they frequently worked in nonagricultural occupations. Yet a far larger proportion of 20-30 years olds from minifundia than on other farms were cacao farmers, rather than workers. also that by age 30-40 few hired-labor-based producers worked off-farm relative to the other groups.

Secondary occupations for minifundia workers tended to be in cacao production off-farm, not discounting those whose own farm work was the secondary occupation. The other farm types were unlikely to work in cacao as a secondary occupation.

Women's formal participation in the EAP was relatively low--under ten percent. When women did work, it was in all types of occupations, although participation in agricultural work varied inversely with size of farm. Over half of active minifundia women were cacao farmers, far more than for other types, while family farm women were most active in unskilled work, and hired-labor-based farm women in agricultural work (at home) and skilled work. There was no particular tendency for younger or older women to predominate in the work force. However, the SST only documented work that lasted for at least 30 days of the year. Field interviews suggested that female participation in cacao harvests, in particular--or in processing--was very common at peak demand periods.

There were only 16 "former cacao farmers" in the sample, who held a variety of occupations. Chart 3.17 shows the 1979 occupations of "former cacao workers," who represented a third of all nonlandholders who had ever been occupied as cacao workers. Nearly a third of the former workers had retired. Remaining minifundistas were doing noncacao agricultural work or skilled trades. Family-labor-based workers were least active in farm work; hired-labor-based workers least active in unskilled work.

Analysis of the historical occupations of cacao household workers showed a general decline in the proportion of cacao farmers and a corresponding increase in agricultural workers and workers in skilled trades and commerce. The proportion of individuals working off-farm in cacao had gone down. The historical occupation of household heads changed less markedly over time, except for minifundistas and collectives. There were relatively more minifundista cacao farmers in 1980 than in 1965; among collective ejidatarios, there was an abrupt occupation change coinciding with the start of the Chontalpa Plan project.

Chart 3.17 Present Occupation of Former Cacao Workers[a,b,c]

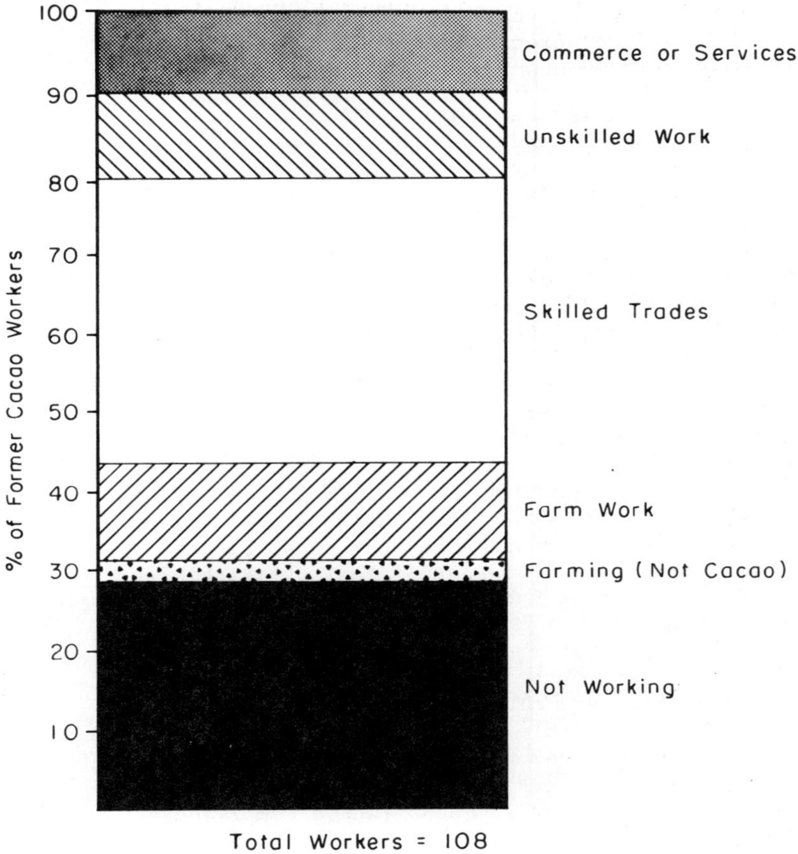

[a]Note that there was one ex-cacao worker on a collective ejido who in 1979 was a farm worker.

[b]This graph does not include cacao farmers who were cacao workers before acquiring their farms, nor cacao workers who once were farmers.

[c]A total of 337 individuals had a history of being cacao farm-workers, without ever owning a farm; "former workers" therefore represent a third of this group.

Source: COPRODET and CONAPO, "Encuesta Sociodemográfica del Estado de Tabasco," Villahermosa, Tabasco, January 1980, Cacao Subsample.

Farm Production and Family Income

The differing assets of the cacao farm types led to production strategies that utilized very different combinations of labor and purchased inputs. Farming intensity and importance of farm income in total family income differed among them.

Chart 3.18 shows some examples from the nine case study farms of per hectare costs of cacao production in Tabasco compared with costs from the recommended farm budgets of the Grijalva Commission. None of the case study farms reached total recommended input levels, even including the imputed value of family labor. The family-labor-based farmers approached the total most closely by substituting family labor for purchased inputs. The exception was a farmer who refused to hire nonfamily labor even for important maintenance activities.

Yield data are summarized in Table 3.5 CCT data suggested that cacao production per tree was slightly higher on minifundia than family farm or hired-labor-based farms. Production per hectare was considerably higher on minifundia, with family farms slightly above hired-labor-based producers. The case study yield data (not statistically significant) showed family-labor-based farm cacao yields to be lower than those on hired-labor-based farms, presumably due to more effective use of labor, either through better sprayers and other equipment or better condition of plantations. The minifundia yields were somewhat lower than the family-labor-based yields, due to the greatly reduced level of purchased inputs of pesticides. The individual overlaps between groups were considerable. The case studies support the contention that management and family characteristics are more important yield determinants than farm size.

For the minifundistas, cacao income comprised a much smaller proportion of total income, so the reduced investment costs made sense. Even for family-labor-based households, cacao income frequently accounted for under half of total income, although total income levels from cacao are much higher than for minifundistas. On the hired-labor-based farms cacao income was extremely important. See Chart 3.19.

The following five chapters provide detailed budget analyses for the case study farms and landless workers, which illustrate and explain the decision-making principles used in the production process and in entering the off-farm labor market. The petroleum economy entered into many of these decisions through changes in off-farm wage levels, inflation in the cost of living and farm input costs, and inflation in the value of land. Through the personal stories of families

Chart 3.18 Per Hectare Costs of Cacao Production on Selected Tabasco Cacao Farms, by Type

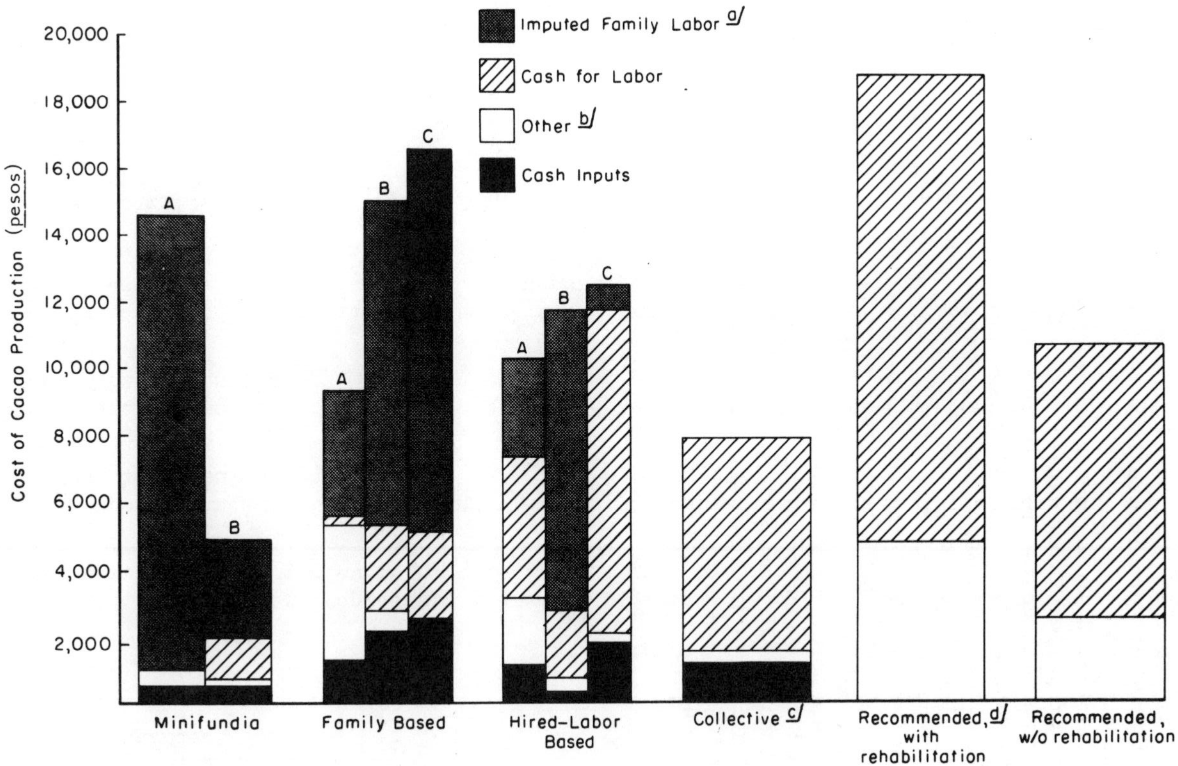

Chart 3.18 continued

Notes

[a]"Imputed family labor" represents the value of family labor used on the farm, had it been remunerated at the going agricultural wage rate.

[b]"Other costs" include transport and vehicle use; depreciation on tools, equipment, and other items; and interest on debt.

[c]Cost data on the collective does not include any depreciation on equipment or buildings, which would be substantial.

[d]The "recommended" production system is taken from the SARH, *Plan de Cacao*, 1979. It does not include any estimates for depreciation, transport, or interest.

Source: Compiled by the author. Tabasco Cacao Farm Studies Case Studies, 1979-81, p. 114.

Table 3.5 Cacao Yields on Tabasco Cacao Farms, by Type

Yield Measure	Minifundia	Family Farm	Hired-Labor-Based
Case Studies[a]			
Production per tree	.28-.45	.35-.77	.42-2.21
Production per hectare	170-301	173-340	254-1,106
Cacao Census[b]			
Production per tree	1.0-1.2	.9	.9
Production per hectare	617-785	510	498
Recommended[c]			
Production per hectare (without rehabilitation	637	637	637
Production per tree (with rehabilitation)	1.0-1.3	1.0-1.3	1.0-1.3
Production per hectare (with rehabilitation)	825	825	825

[a]Author, Tabasco Cacao Farm Case Studies, 1979-81.

[b]SARH, "Censo de Cacao," Villahermosa, Tabasco, 1978.

[c]SARH, Representación Tabasco, "Plan de Incremento de la Producción de Cacao en el Estado de Tabasco," 1980.

Chart 3.19 Family Income on Tabasco Cacao Farms, by Source and Farm Type

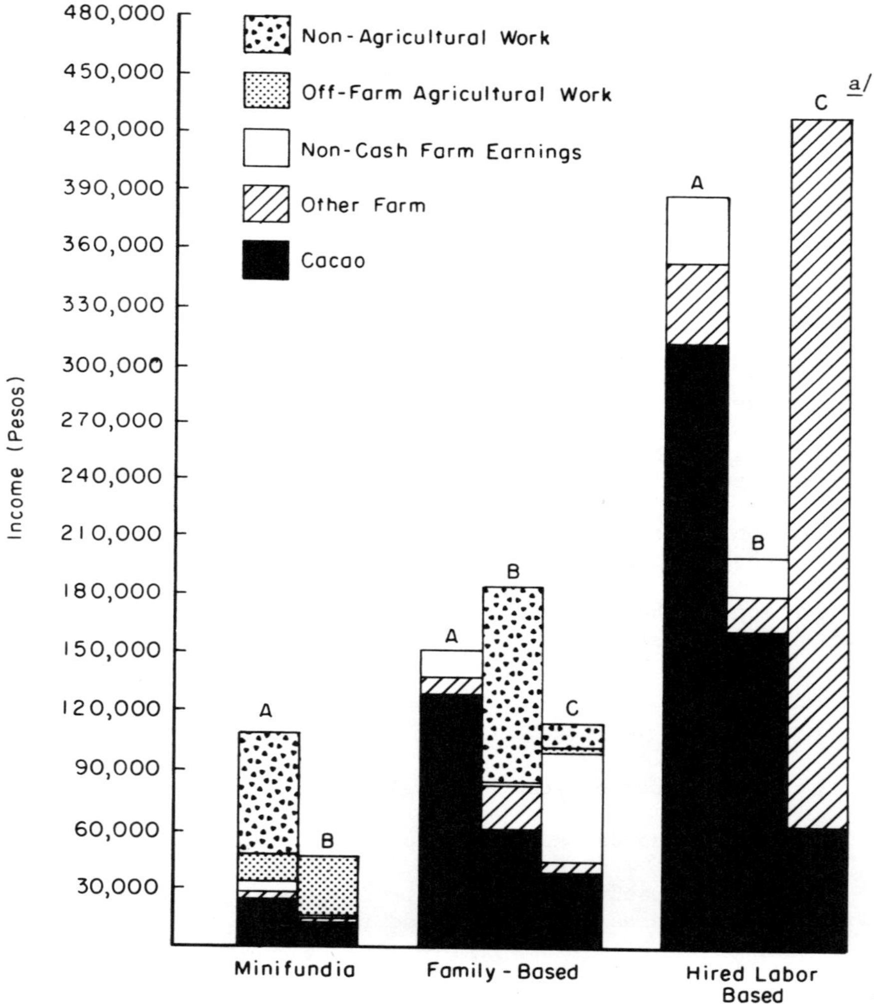

[a]Multiply all figures by 10 for this family.

Source: Author, Tabasco Cacao Farm Case Studies, 1979-81.

trying in their individual ways to protect and promote the family welfare--and the statistical story of many families organized like them--those chapters will analyze the family economies of Tabasco's cacaoteros.

4

Hired-Labor-Based Farm Production Strategy

It is unlikely that during the nineteenth century many farms in the campesino sector in Tabasco hired "free labor," given a generally nonmonetary economy, tight community organization, and traditions of labor exchange. Thus the extent of campesino holdings were limited by family size and composition. The larger plantations set up by descendents of the Spanish were predominantly based on hired labor or debt peonage. Workers were usually permanently employed; where extra labor was required at harvest time, indigenous workers were hired. The owners themselves performed very little of the actual work; rather, they organized and supervised work, though this was often left to overseers. During the period when the Sierra dominated production, these "largeholders" probably produced a much larger percentage of the state's cacao than is the case today.

This situation underwent basic structural change during the last few decades; as the cacao frontier disappeared, active markets in labor and land developed, and many largeholders changed to cattle activities. The appearance of large numbers of landless workers and minifundistas who needed part-time work off their own cacao farms coincided with the growing use by campesinos of hired labor. With the exception of a few very large latifundia-type farms, most cacao farms were relatively small in area, rarely over 20 hectares. Many of these farmers had close family ties with campesino producers and had become HLB producers via a gradual accumulation of resources from a campesino base.

The Survey results summarized in Chart 4.1 show that hired-labor-based cacao producers were mainly older farmers (their average age was 55), with large families and children of working age. For many campesinos, the large farm was the result of a lifetime of consolidation efforts and partly made possible (as well as necessary) by a large number of dependents who could help in farm work as the farm was developed. Most farmers received any land as inheritance decades ago when the average plot size was larger than it is today.

These farm entities were not stable over time. Land inheritance, whether taking place during the lifetime of the farmer or after his death, often broke up these cacao plantations into family or minifundia type holdings. For the more clearly capitalist types, accumulation was often the result of attempts to accumulate capital earned outside agriculture in cacao plantations and thereby to establish status in an agrarian culture as well as a profitable business.

MEASURING FARM COSTS AND FAMILY INCOME

Analysis of economic decision making on mixed commercial-subsistence farms is complicated by problems of definition of costs, income, and profits, as discussed in Appendix A. In this research, I resolved the problems by collecting disaggregated data on cash, real, and financial costs and returns of production and then comparing the resulting budgets with farmers' assertions about their decision-making strategies and priorities. This system was used on a limited basis with the regional cacao farm surveys, and in-depth information was collected for the cacao farm case studies.

Below are the relevant definitions and guidelines for budgetary analysis and selected data tables for the three case study hired-labor-based farms. More extensive data on farm assets, farm labor use by type of worker and farm activity, family labor use on- and off-farm, household expenditures, livestock management, and monthly income flows may be found in Scherr (1983b). Tables 4.1 and 4.2 show cacao inputs and production costs and Table 4.3, all farm production costs. All financial information is in 1980 pesos. Because land prices were so important in budget analyses, all calculations were made for both high and low estimates for land value. Depreciation costs were assigned to a specific enterprise or, where asset use was shared, equally between enterprises. Remember that figures presented in dollar equivalents in fact, overstate dollar value, as the peso was highly overvalued during this period.

Costs were measured in three ways to permit analysis from various points of view. The first was "cash costs," which included only those inputs for which cash was paid and did not include family labor. The second, "full costs," included depreciation and family labor, which was valued at the typical (not minimum) going rate of $120 per day. The third, "costs with equity deduction," added an arbitrary (low) capital cost

Chart 4.1 Demographic and Employment Characteristics of Hired-Labor-Based Cacao Farm Households

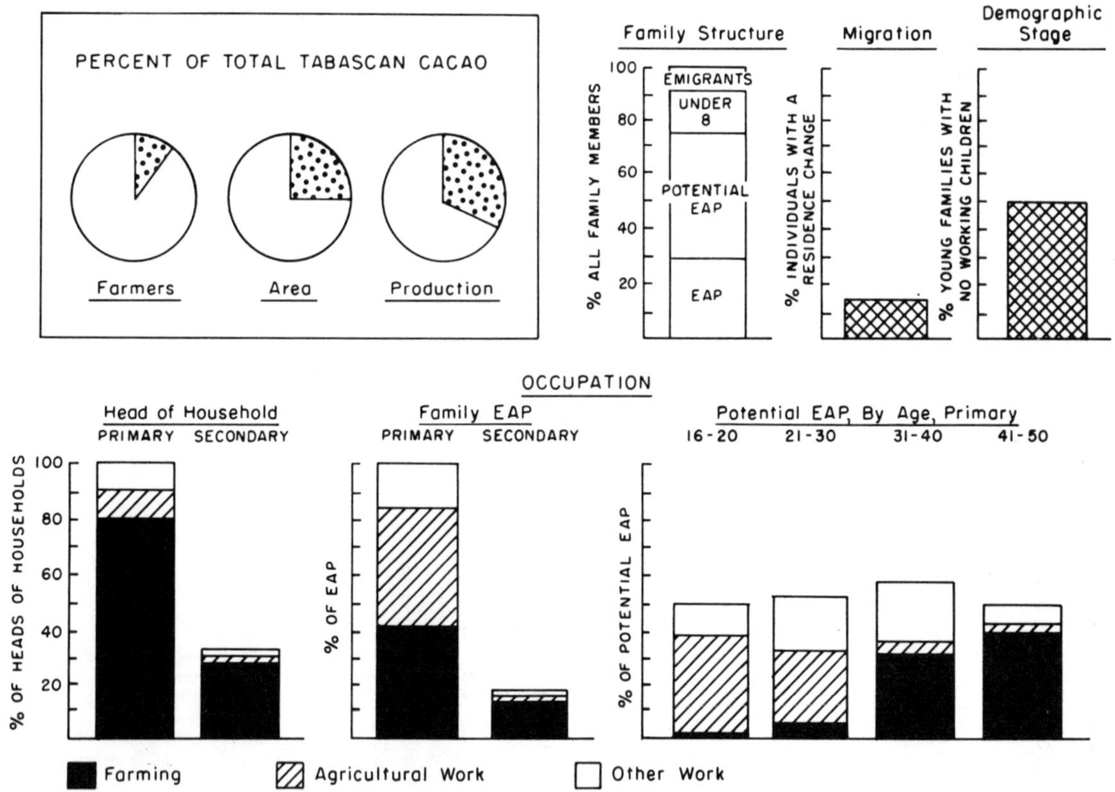

Table 4.1 Inputs for Hired-Labor-Based Cacao Farms, 1978-80

	Prado Farm	Ramirez Farm	Gomez Farm	*Recommended*[a] w/o Rehab	*Recommended*[a] with Rehab
Productivity					
No. trees	24,000 (600/h)	3,000 (662/h)	5,300 (662/h)	600/h	600/h
Age 1	40 yrs-14,400	4 yrs-375	15 yrs-2,800		
Age 2	10-20 yr-9,600	15 yrs-2,625	10 yrs-2,000		
Age 3	--	--	6 yrs-500		
Pdn/tree	.424 kg	2.21 kg	.6 kg	Average 1.3 kg	
Pdn/hectare	254.5 ($12,750)	1,106 kg ($55,300)	395 kg ($19,750)	Average 538 kg	
Man-days/hectare	82.5-87.5	52.6-60.6	61.9-76.7	*81.5*	*130*
Pruning/clearing	20 (overestimated)	18.0	2.0	21.0	24
Drains	5.5		3.7	3.0	6
Fertilizer		3.0	3.4	2.5	10
Insecticide	14	6.3	6.9	12.0	20
Fungicide		5.3	7.7	15.0	30
Harvest & pod-breaking	43-48	20-28	38.2-53	28.0	40
Inputs/hectare					
Fertilizer					
Urea	-- --	1 (with metílico)	--	--	2
18-9-18	-- --	1	--	--	--
15-15-23	1	--	--	--	--
Floren	-- --	--	1	--	1-2

Table 4.1 Continued

	Prado Farm	Ramirez Farm	Gomez Farm	*Recommended* w/o Rehab	*Recommended* with Rehab
Inputs/hectare					
Fertilizer *(cont'd)*					
Nutrisol	--	2	--	--	--
Gro-Green	--	(with metílico)	--	--	--
Crescal	--	--	--	--	--
Insecticides			--		
Trident	--	2	--	--	--
Metilico	--	2	2	--	2-3
Foley	--	--	2	--	--
Folidol	2	--	--	--	--
Folimac	--	--	--	--	--
Polidor	--	--	--	--	--
Fungicide					
Lime & sulfate	[formerly 10]	4	3	--	5-8
Cuprasol	-- --	--	--	--	--
Pruning	[formerly 5-6]	3-4	2	--	2-3

Note: kg = kilogram; h = hectare.

[a]This data is from estimates in the SARH, *Plan de Cacao*, 1979.

[b]This data is from the SARH 1978 Cacao Census.

Source: Author, Cacao Farm Budget Studies, 1979-81, Tabasco, Mexico.

Table 4.2 Costs of Cacao Production on Hired-Labor-Based Cacao Farms, 1979-80

Inputs	Prado Farm	Ramirez Farm	Gomez Farm	Recommended[a] Plan Grijalva[b]	w/o Rehab[c]	with Rehab[d]
LABOR	$ 390,000	$ 25,920	$ 14,575	$ 9,400	$ 8,100	$ 13,900
Interest	—	—	—	—	—	—
INPUTS	75,600	9,800	3,715	5,075	2,547	4,731
Miscellaneous	24,500	2,800	1,525	--	--	--
Fertilizer	19,000	2,680	250	3,200	1,500	2,366
Urea	--	80	--	--	--	--
18-9-18	19,000	2,200	--	--	--	--
Floren	--	--	250	--	--	--
Gro-Green	--	--	--	--	--	--
Nutrisol	--	400	--	--	--	--
Insecticide	12,600	2,120	1,400	675	1,047	2,365
Folimac	--	--	--	--	--	--
Tridente	--	1,280	--	--	--	--
Metilico	--	--	--	--	--	--
Methyl	--	840	1,000	--	--	--
BHC	--	--	--	--	--	--
Foley	--	--	400	--	--	--
Folidor	--	--	--	--	--	--
Folidol	12,600	--	--	--	--	--
Fungicide	19,500	2,200	540	1,200	--	--
Lime	1,500	200	--	--	--	--
Sulfate	18,000	2,000	540	--	--	--

Table 4.2 Continued

Inputs	Prado Farm	Ramirez Farm	Gomez Farm	Recommended Plan[a] Grijalva[b]	v/o Rehab[c]	with Rehab[d]
CASH COSTS	$ 465,600	$ 36,020	$ 18,290	$ 14,475	$ 10,547	$ 18,631
DEPRECIATION	11,900	9,145	800	--	--	--
FULL COSTS	477,500	45,165	19,090	14,475	10,647	18,631
EQUITY DEDUCTED[e]	563,115 (355,155)	90,457 (48,457)	112,560 (56,560)	14,000 (10,500)	14,000 (10,500)	14,000 (10,500)
COSTS WITH EQUITY Deduction	(830,615) 1,040,615	135,622 (93,622)	131,650 (75,650)	28,475 (24,975)	24,647 (21,147)	32,631 (29,131)
Number of Hectares	40	6	7	1	1	1
TOTAL CASH COSTS	11,640	6,003	2,612	--	10,647	18,631
Family man-days/hectare	5.2	23.5	69.0	--	--	--
Imputed Cost/family labor	624	2,820	8,280	--	--	--
CASH FOR INPUTS	1,890	1,166	530	--	2,547	4,731
WITH DEPRECIATION[f]	11,937	7,526	2,727	--	10,647	18,631
WITH EQUITY DEDUCTED[f]	26,015	22,603	18,807	--	24,647	32,631

Notes:

[a]The original figures valued a man-day of labor at $75 (Grijalva data) and $83 (Cacao Plan data). Here we have adjusted this to $100, the minimal going rate found in Tabasco during my field study. At the more typical rate of $120, the estimates would have been substantially higher.

[b]SARH, Comisión de Grijalva, Dirección de Estudios y Proyectos, "Rendimientos y Costos de Producción de Cultivos Tropicales," August 1979, pp. 10-16.

[c]SARH, "Plan de Incremento de la Producción de Cacao en el Estado de Tabasco," January 1979, Villahermosa, Tabasco, Cuadro 20. This represents the supposed "typical" production system in Tabasco before rehabilitation programs.

[d]Same source as above. This represents the production costs after farmers have instituted rehabilitative measures on their plantations (see text) suggested by the Cacao Plan.

[e]A higher estimated value for land is shown in the first figure; a lower estimate in parentheses.

[f]These include imputed cost of family labor.

Table 4.3 All-Farm Production Costs on Hired-Labor-Based Cacao Farms, 1979-80

Inputs	Prado Farm	Ramirez Farm	Gomez Farm
Labor	$ 965,000	$ 33,600	$ 17,250
Cacao	--	25,920	14,575
Cattle	--	7,680	1,600
Maize	--	--	--
Cacao chemicals & miscellaneous	39,600	7,000	2,715
Vehicle[a]	133,000	1,200	1,500
Cattle Feed, etc.	755,000	1,800	5,460
Veterinary services & medicine	60,000	--	2,000
Equipment	--	--	--
Cattle purchases	50,000	--	--
Interest on loans	170,000	14,000	--
Rent, land, equipment	177,000	5,300	--
Miscellaneous	--	--	1,125
CASH COSTS	1,405,600	62,900	30,050

Depreciation	288,000	37,500	4,100
Cacao	11,000	9,145	800
Cattle	--	18,366	300
Maize, etc.	--	--	--
Presents of products	--	12,000	--
FULL COSTS	1,693,600	112,400	34,150
Equity capital costs	1,050,000-1,775,795	210,207-294,557	86,800-147,770
Cacao	353,115- 563,115	48,457- 90,457	56,560-112,560
Cattle:	747,425-1,044,925	117,450-159,800	6,860- 8,960
Pigs	27,055- 34,055	--	--
Other animals	481,000	--	--
Maize, etc.	--	--	--
COSTS with deduction of 7% on equity capital	2,743,600-3,469,395	312,607-396,957	120,950-181,920

[a]This value includes tractors, trucks, etc.

Source: Author, Cacao Farm Budget Studies, 1979-81, Tabasco, Mexico.

Table 4.4 Hired-Labor-Based Cacao Farm Household Gross Income, 1979-80

Source	Prado Family	Ramirez Family	Gomez Family
Cacao	$ 626,728 (14.6%)	$311,070 (80.0%)	$162,156 (80.4%)
Poultry	--	--	--
Garden	--	--	--
Cattle	120,000 (6 bulls @ $20,000) 165,000 (50 cows @ 150 kg; @ $22 kg) 975,000 (150 calves @ 6,500) 140,000 (20 heifers @ 7,000) $1,400,000	35,000	10,000
Rent pasture	--	--	--
Pig sales	1,944,000 (600 @ 90 kg @ $36/kg)	5,000	1,750
Cattle sales	--	--	--
Milk	120,000	--	--
Pepper	--	--	--
Bananas	--	--	--
Zapote	7,000	--	--
Wood	120,000	--	--
Coconut	--	--	5,000
Oranges	--	1,600	--

Car rental	--	--	1,250
Miscellaneous	--	--	--
TOTAL FARM CASH INCOME	$4,289,728	$352,670	$180,156
Farm produce consumed			
Products	50,000	1,000	
House	150,000	--	10,000
Pigs	--	--	500
Cacao, coconut, banana	--	--	1,000
	200,000	1,000	11,500
Change in livestock			
inventory: Pigs	+15,000		
Cattle	+200,000	+37,500	+10,000
	-500,000	- 2,500	
	-285,000	35,000	10,000
Miscellaneous	--	--	--
TOTAL FARM LABOR EARNINGS	4,274,728	388,670	201,656
Off-farm agricultural work	--	--	--
Off-farm nonagricultural work	--	--	--
Loans	--	--	--
TOTAL INCOME	4,274,728	388,670	201,656

Source: Author, Cacao Farm Budget Studies, 1979-81, Tabasco, Mexico.

of seven percent of assets to represent the income which could have been earned had all assets been sold and the proceeds invested in the bank.

Table 4.2 shows total cacao production costs; it analyzes per hectare costs to permit comparison with other farms and with "recommended" budgets. This extracts the cash costs for inputs and the imputed cost of family labor (at the going wage rate for agricultural labor), then recomputes per hectare cash costs, full costs (including depreciation and imputed family labor), and costs with equity deduction (which also includes imputed family labor costs).

Gross household income, shown in Table 4.4, was measured in three ways: "cash income," the cash value of farm products marketed; "total farm labor earnings," which included an imputed value of farm produce consumed and any change in the livestock inventory; and "total income" which added to these any income from off-farm sources. Monthly income flow data was also collected to determine when during the year decision making might be the result not of overall financial considerations but rather cash flow problems. All current income was agriculture-related for these three families. The Prado and Ramirez families had formerly had major commercial and urban investments. At the time of the interview they were considering other such investments.

Net family and farm income were measured by a number of alternative methods. The principal measures show net cash income, labor income, labor earnings disregarding invisible equity costs, and labor earnings after deduction of equity costs. Income measures are provided for the principal farm enterprises and off-farm income sources. Then the average value of labor in the enterprise was computed--for *all* labor used (i.e., an implicit wage, equivalent to average, not marginal costs), and specifically for *family* labor used. These estimates were tested for sensitivity to data errors of 20 percent in cost and labor use estimates. These were used to compare income-earning alternatives for the families between farm enterprises and between farm and nonfarm income sources. The final variables reported are the proportion of labor costs in cacao and whole farm expenses.

Information on household expenditure for the three families was also collected, although it is probably the least reliable of all the data (see Scherr 1983b). The figures represented only a year's expenditures, whereas certain large purchases made in other years should have been depreciated--for example, furniture and consumer durables. Expenditures fluctuated greatly with yearly income level, and many high-priced items were purchased only from unexpected income surpluses.

TYPICAL HLB FARM HOUSEHOLDS: CASE STUDIES

This section will present in-depth data and analysis for three hired-labor-based families. They provide examples of the economic conditions and responses of the major kinds of hired-labor-based farmers--the wealthy largeholder, the recent urban investor in cacao farming, and the campesino who amassed a large number of plots over time. Sr. Prado* was an ambitious largeholder and rancher from Teapa; Sr. Ramirez was a very urban and enterprising cacaotero from Huimango town; and Sr. Gomez was a barely literate campesino who, through good management and inheritance, acquired a large farming base.

Prado Family

Sr. Prado was born and raised in Teapa. He rose from peon to wealthy politician through his work on the land. He was 59 and had a wife and four children of whom three were college trained. None lived at home.

Landless until 1949, he worked two untitled lots of ten hectares each, which were later enlarged and made part of an ejido. He bought additional land in 1960, 1962, 1964, and 1971, all on credit. In 1980 he owned 40 hectares of cacao and 250 hectares of pasture. His assets--all agricultural--were worth over US$1 million, with over three-fourths in land, the rest in buildings and equipment. He owned 39 adult pigs and their offspring, and a cattle herd of 500. During the agricultural year 1979-80, his cacao plantations produced 10,179.8 kilos of dried cacao beans.

Sr. Prado worked independently, although one son helped him with cattle activities, and he was helping another son invest in a hog operation. But he had 27 permanent year-round workers, of whom 12 were occupied in cacao. Of these, 14 were over fifty years old and had been working for Sr. Prado for a long time. Due to feelings of obligation, he retained several who were no longer very productive. Half lived on his property in houses constructed by him. Only the older workers preferred "resident" status; younger workers preferred freedom to find the best-paid work. Of total labor used on the farm (approximately 8,945 man-days), less than three percent was carried out by Sr. Prado, five percent by his family, and over 92 percent by hired workers.

Full-time wages varied considerably from $40 for a young person for milking cows twice a day, to $700/week ($116/day) plus $10 per pig produced, for the pig manager. Part-time workers were paid $70/half-day and $120/day to work 8:00 A.M.

---

*Please note that all names of case study farmers have been changed to protect their privacy.

Table 4.5A Family and Farm Income of Hired-Labor-Based Tabasco Cacao Farms, 1979-80: Prado Farm

Income Measure[a]	Cash Income[b]	*Results* Labor Earnings[c] without Equity Return Deduction	Labor Earnings[d] with Equity Return Deduction
a. Net farm ___	$1,939,128.00	$1,446,128.00	$ -329,667.00
b. Net off-farm agricultural ___	--	--	--
c. Net nonagricultural ___	--	--	--
d. Net total ___[e]	1,939,128.00	1,446,128.00	-329,667.00
e. Net farm ___ per hectare	6,780.17	5,056.39	-1,152.68
f. Net farm ___ per man-day on farm[f]	311.25	161.67	-71.03
g. Net farm ___ per family man-day on farm[g]	3,006.40	2,242.06	-511.11
h. Net nonagricultural ___ per family man-day worked outside agriculture[g]	--	--	--
i. Net agricultural off-farm ___ per family man-day worked[g]	--	--	--
j. Net total ___ per family man-day worked[g]	3,006.40	2,242.06	-511.11
k. Net cacao ___ per man-day in cacao[f]	144.48	123.85	21.82 (38.86)[h]
l. Net cattle ___ per man-day in cattle[f]	145.97	86.43	92.72 (-202.71)[h]
m. Net pig ___ per man-day in pig work[f]	3,967.66	4,102.34	4,035.00 (3,963.00)[h]

n. Net __ ___ per man-day in __[f]	--	--	--
o. Net cacao __ per family man-day in cacao[g]	497.86	174.79	-1,423.02
p. Net cattle __ per family man-day in cattle[g]	956.00	112.36	-2,151.00
q. Net pig __ per family man-day in pigs[g]	30,900.00	32,382.80	30,841.00 (30,356.00)
r. Net __ ___ per family man-day in __[g]	--	--	--
s. Percent labor in cacao expenses[h/j]	81	82	47
t. Percent labor in whole farm expenses[h/j]	36	32	23

Note: Irregularities in the Prado budget include: a) unusual reduction in the cattle inventory this year; b) milk income estimates which seem too low; c) labor use in pig operation probably underestimated; d) no payments of principal on loans are included as costs; e) it is not certain whether the owner's son received cash income for work in the cattle operation--none is included here; f) labor input of Sr. Prado is probably underestimated.

[a]The blank spaces in this column refer to the specific income measure used--cash and labor earnings with and without equity deduction; the figures are found in the appropriate column to the right.

[b]Cash income refers to costs and returns to the family only in cash.

[c]"Labor Earnings" (without equity deduction) is a measure of income that includes several noncash components such as inventory changes, family use of farm products, etc.

[d]"Labor Earnings" with equity deduction is similar to the simple labor earnings measure, but it also adds the cost of tying up the family's capital in the farm enterprise. It enters, as a cost, the

Table 4.5A Continued

interest that could be paid to the family by a bank were all assets to be sold for cash and deposited, without the risks entailed in a farm enterprise. The very low rate of 7 percent is used for our calculations.

[e]Total Income of whatever measure is the sum of farm income, off-farm agricultural income, and off-farm nonagricultural income.

[f]Income per man-day is a rough measure of average labor productivity including both hired and family labor use. To compute this, the wage bill is subtracted from total costs.

[g]This is the average value of a day of labor performed by family members. For this purpose the wage bill *is* included in costs.

[h]The value in parentheses represents the higher of two cost estimates for cacao production.

[i]This pig budget is probably inaccurate. If we assume net returns at $1 million (based on unreported costs) and a doubling of the very low reported labor input to 800 man-days per year, then returns per man-day are reduced to $1,250.

[j]This does not include the imputed value of family labor.

[k]The value in parentheses represents costs (or returns) in the form of a changed animal inventory.

Source: Author, Cacao Farm Budget Analyses, 1979-80, Tabasco, Mexico. Note that the costs, returns, and labor use summarized here can be found in preceding chapter tables.

50 4:00 P.M. This was the going rate for Teapa fieldworkers, but rather low for cacao harvest work.

Sr. Prado's priority until very recently was cacao, but he had been steadily expanding cattle operations and was putting most of his resources into the pig operation. It is very hard to interpret his labor use. My statistics only included labor hired, and many recorded man-days were of dubious productivity. He had "abandoned" his cacao plantations due to high labor costs and in 1979-80 applied very few inputs. Well over half of all farm production cash costs went to cattle enterprise expenses.

During the prior 20 years cacao production varied widely. When prices were very low in the early 1960s, Sr. Prado renovated most of his plantations. In the mid-1960s he only harvested and gave minimum maintenance to his cacao, emphasizing banana production then. He also sold a lot of cacao land to build up his cattle herd. But he cut down ten hectares of bananas in the early 1970s to replace with cacao as prices rose. He increased pest control and new plantings, reaching maximum production in 1978, but since then pig production, labor problems, and lower cacao prices had led to decreased attention to cacao, mainly maintenance operations to prevent infestation or deterioration of the plantations. Should prices go up again sharply, he might try to set up an irrigation system and renew extensive use of chemical inputs.

Table 4.5A shows Prado family and farm income measures. "Cash income" per family man-day worked was $3,006, with $497 per family man-day in cacao, versus $956 in cattle and $30,900 for pig farming. Inventory changes in cattle, non-cash payments in workers, and depreciation caused "labor earnings" to decline--to $174 per family man-day overall. Adjusted for equity return, this gave a negative farm income and a loss of over $1,000 per hectare per year. Only in pig production were returns to family labor positive.

The main constraint on cacao production was difficulty finding dependable labor, almost certainly a result of his reluctance to pay a "competitive" wage. This was so, even though it would almost certainly be profitable to hire more, even at the higher wage, and even retaining the old workers. The principal reason that he did not do so was that the hog operation was so very much more profitable, without requiring the always difficult labor management of cacao.

Sr. Prado considered that his most important inheritance to his children was their education. One son was interested in the cattle business and one in the pig operation, but neither cared about the cacao--a personal disappointment, as their father had been highly active in cacao politics. This succession problem was probably an important reason why the plantations had been abandoned.

Table 4.5B Family and Farm Income of Hired-Labor-Based Tabasco Cacao Farms, 1979-80: Ramirez Farm

Income Measure[a]	Cash Income[b]	*Results* Labor Earnings[c] without Equity Return Deduction	Labor Earnings[d] with Equity Return Deduction
a. Net farm __	$289,770	$276,270	$ -8,289 (79,063)
b. Net off-farm agricultural __	--	--	--
c. Net nonagricultural __	--	--	--
d. Net total __ [e]	289,770	276,270	-8,289 (76,063)
e. Net farm __ per hectare	6,036	5,755	-172 (1,584)
f. Net farm __ per man-day on farm[f]	573	549	44 (194)
g. Net farm __ per family man-day on farm[g]	988	942	-28 (259)
h. Net nonagricultural __ per family man-day worked outside agriculture[g]	--	--	--
i. Net agricultural off-farm __ per family man-day worked[g]	--	--	--
j. Net total __ per family man-day worked[g]	988	942	-28 (259)
k. Net cacao __ per man-day in cacao[f]	685	664	458 (554)
l. Net cattle __ per man-day in cattle[f]	180	396	-1,128 (-1,678)
m. Net pig __ per man-day in pig work[f]	--	--	--

n. Net ___ ___ per man-day in ___[f]	--	--	--
o. Net cacao ___ per family man-day in cacao[g]	1,185	1.146	756 (937)
p. Net cattle ___ per family man-day in cattle[g]	478	1,758	-7,276 (-10,534)
q. Net pig ___ per family man-day in pigs[g]	--	--	--
r. Net ___ ___ per family man-day in ___[g]	--	--	--
s. Percent labor in cacao expenses[h/j]	71	57	19-27
t. Percent labor in whole farm expenses[h/j]	53	21	8-10

Note: Irregularities in the Ramirez budget include: a) underestimation of fungicide applications; b) no payment of principal on loans is included as costs.

[a]For footnotes, see Table 4.5A, pp. 133-34.

Source: Author, Cacao Farm Budget Analyses, 1979-80, Tabasco, Mexico. Note that the costs, returns, and labor use summarized here can be found in preceding chapter tables.

Ramirez Family

Sr. Ramirez, 54, was born in Cunduacán town where he worked in accounting and business. At age 22, he and his wife moved to rural Huimango I to open a store, the proceeds from which were invested seven years later in two hectares of cacao land. Ten years later he gave up the store to concentrate on cacao production and ranching, after adding 18 hectares of pastureland and his wife's inheritance of four hectares of cacao. In 1978, he bought another 24-hectare ranch. The couple had 11 children, of whom three were married and in urban occupations; three were in professional schools; and five were school-age.

The Ramirez assets were estimated to be worth around US$130,000-180,000, about three quarters in land. Total cacao production for 1979/80 was approximately 6,636 kilos of dried beans. The family cattle herd included 67 animals. There were no longer any significant off-farm assets. Sr. Ramirez' farm supported ten people at an average expenditure of $31,568 per person (two to three times subsistence levels). Forty-four percent of the family budget went for food (clothes were not included because almost none were bought that year); educational expenses were also very high.

Sr. Ramirez worked without any full-time assistance, despite time-consuming involvement in community and cacao politics. For four months of the year the farm required less than one full-time worker. During May, July, October, November, and April two were required for at least part of the month. From January through March the farm was a four-worker operation; 50 percent of all labor was performed at this time. Forty-two percent of all farm labor was related to cacao harvest or postharvest operations, for which workers were hired. Cattle production accounted for about 14 percent of all labor utilized on the farm.

All family labor, except for the wife's minimal participation in pod-breaking, took place in the afternoons or on weekends, secondary to other activities. Of total labor used on the farm during 1979-80 (564 man-days), 27 percent was done by Sr. Ramirez, 25 percent by his family, and 48 percent by hired labor. A very large number of individual laborers were used (20), most of whom returned yearly and were paid a relatively high wage ($20 per hour at the lowest level). Relations between Sr. Ramirez and his workers were the most "businesslike" of the three farmers.

He also fertilized and used insecticides more heavily. The Ramirez farm came fairly close to fulfilling Plan recommendations (see Table 4.5B). Net cash farm earnings were high: $1,699 per family man-day (US$75) versus $1,003 overall on the farm and $478 in cattle. Labor earnings were $977 per family man-day; $1,644 in cacao, $1,758 in cattle. With equity included, the average family man-day earned between \$27 and \$315 overall--\$1,099-1,352 for the cacao enterprise. The cattle operation showed a loss, mainly because building up the herd permitted few sales. Sr. Ramirez' main temporary constraint was lack of capital, for he went into debt to finance the expanded cattle operation. He expected that finance would not be a problem after cattle sales began in earnest.

When the college-age boys were out of school, only two sons would remain to help out part-time. Like Sr. Prado, Sr., Ramirez considered that his chief inheritance would be his children's education. Indeed, none seemed interested in a career farming the land. While the father was interested in buying more land, the lack of labor and of children's interest restrained him.

Although Sr. Ramirez was not yet overly concerned about the labor problem, it would seem that hiring a full-time assistant would be worthwhile. Even with social security, an assistant's pay would not be above \$42,000 per year; extra harvest labor would also be needed in January-March, assuming higher yields. Total costs would increase by \$20,000, but by so doing, care could be provided to cattle at calving, and cattle and cacao would not compete for attention during the cacao harvest. It would take only the marketing of two extra steer per year or an increase in cacao production of 200 kilos (three percent) to justify the expense. The family could also consider renting pasture in the dry months to supplement their own, thereby increasing the carrying capacity of their land.

His historic response to cacao conditions suggest that Sr. Ramirez was very sensitive to issues of alternative investments and profits. He acquired his cacao land in 1955 and 1962; since then he had concentrated on increasing production by renovation. He replanted three hectares around 1965 and another in 1976. He says he produced less during the 1960s, but that goods were less expensive, and he supplemented production with income from the store.

As cacao prices rose, he got rid of the store; in 1975 he ceased maize production. He asserted that if consumer prices continued to rise and cacao prices to fall, he would seriously consider opening a modern butcher shop for which he was already drawing up plans. If the cacao price were to go as low as \$30 per kilo, he would abandon his plantation. If cacao prices increased, he planned to add more chemicals, do more careful pruning, and renovate his drains.

Gomez Farm

Sr. Gomez was 78, with no formal schooling. He was the son of one of the first campesinos to settle in Huimango II.

Table 4.5C Family and Farm Income of Hired-Labor-Based Tabasco Cacao Farms, 1979-80: Gómez Farm

Income Measure[a]	Cash Income[b]	*Results* Labor Earnings[c] without Equity Return Deduction	Labor Earnings[d] with Equity Return Deduction
a. Net farm __	$137,586.00	$151,286.00	$50,786.00 (-10.18)[h]
b. Net off-farm agricultural __	--	--	--
c. Net nonagricultural __	1,250.00	1,250.00	1,250.00
d. Net total __[e]	138,836.00	152,536.00	52,036.00
e. Net farm __ per hectare	15,287.00	16,809.00	5,642.00 (-1,131)[h]
f. Net farm __ per man-day on farm[f]	258.51	222.94	78.28 (-23.34)[h]
g. Net farm __ per family man-day on farm[g]	354.14	365.43	122.67 (-24.60)[h]
h. Net nonagricultural __ per family man-day worked outside agriculture[g]	156.25	156.25	156.25
i. Net agricultural off-farm __ per family man-day worked[g]	--	--	--
j. Net total __ per family man-day worked[g]	329.00	361.46	123.31 (-21.17)[h]
k. Net cacao __ per man-day in cacao[f]	272.37	271.63	167.86 (65.10)[h]
l. Net cattle __ per man-day in cattle[f]	1.87	211.27	99.01 (52.86)[h/k]
m. Net pig __ per man-day in pig work[f]	--	--	--

n. Net ___ per man-day in ___ [f]	--	--	--
o. Net cacao __ per family man-day in cacao[g]	354.14	353.08	203.46 (55.31)[j/h]
p. Net cattle __ per family man-day in cattle[g]	-46.81	222.64	80.69 (22.36)[h/k]
q. Net pig __ per family man-day in pigs[g]	--	--	--
r. Net __ ___ per family man-day in __ [g]	--	--	--
s. Percent labor in cacao expenses[k/j]	52	50	17.0 (10)[h/i]
t. Percent labor in whole farm expenses[k/j]	42	36	13.7 (9.3)[h/i]

[a]For footnotes, see Table 4.5A, pp. 133-34.

Source: Author, Cacao Farm Budget Studies, 1979-80, Tabasco, Mexico. Note that the costs, returns, and labor use summarized here can be found in preceding chapter tables.

While single and a young married man, he worked primarily as a jornalero. When he was forty, he received his share of his father's land, eight hectares, of which he sold the four hectares of pastureland. When he married, his in-laws gave one hectare and later sold eight hectares to him. He substituted Guayaquil for Criollo on the cacao land he received earliest. After 1,500 of those trees began to produce in the late 1960s, he ceased maize production altogether to concentrate on cacao. His production steadily increased over the next ten years, although he began to produce maize again when cacao prices plummetted in the mid-1960s. In 1972 he bought another one-hectare parcel and sold it in 1978.

Sr. Gomez and his wife had six children, none living at home. The four daughters married young and were housewives. Both sons worked in cacao--the youngest, Antonio, with his father. None had formal schooling past the third grade.

Sr. Gomez' assets were worth nearly US$90,000, nearly all in land. He had a small herd of cattle and assorted small livestock. Total cacao production for 1979-80 amounted to 3,162 kilos of dried beans. His farm required one man May through August; two men September and January; three men in March and April; four men in December; and five full-time workers in the peak harvest months of October-November. Of total man-days worked on the farm (692), 25 percent were by Sr. Gomez, 45 percent by his family, and 30 percent by hired workers.

Only Antonio was fully employed on the farm. Sr. Gomez was semiretired; he worked only two-thirds of the year, and he admitted that his productivity was only half that of his son. Yet only in October, November, and March did Antonio work a longer-than-usual (25-day) month--under pressure. For five months of the year, he worked less than 15 days per month. In addition to farm work he rented out their truck about eight times a year for trips to Cunduacán and Comalcalco.

To meet labor needs Sr. Gomez hired 13 different men, half of whom had some kinship or social relation to him. One older man who worked on drains, his compadre, was offered work more out of charity than necessity. His wage was approximately $16 per hour for a longer day. For drain work or pasture weeding the wage was much higher but paid for a completed job, not by the hour. During 1980-81, when yields went down significantly due to the weather, the amount of labor hired dropped, despite the fact that the father's worsening physical health meant that his participation stopped. Sr. Gomez felt he had no short-term flexibility about wages or labor hired at harvest; he absolutely had to have help, so he paid the price.

Sr. Gomez' cash income in 1979-80 was $137,586, or $332 per family man-day worked, with the cattle enterprise operating at a loss (see Table 4.5C, pp. 140-41). Labor earnings were about $365 per family man-day, and cattle enterprise returns were positive. With a low equity return deduction estimate, returns were $123 per family man-day--barely the going minimum wage--although the cacao enterprise earned $203 per family man-day. Using the high equity estimate, returns were negative overall and only $55 per family man-day in cacao. These negative values were mostly attributable to four hectares of land "underutilized" in maize, and high costs for an expensive new house and truck.

For Sr. Gomez, labor was a significant constraint, because only one full-time worker was available for a plantation which half the year required over three full-time people. The family wanted to find a permanent worker but was unwilling to pay the necessary price. They would still need part-time labor at harvest, and the man would be extraneous for four months. As it was, they could not find "reliable" labor. For a total of under one man-equivalent, 13 workers were hired during the year, half of whom had never worked there before. Given the very high returns to labor in this operation, it would make good economic sense for Sr. Gomez to offer higher-than-competitive wages to attract steadier workers.

Sr. Gomez claimed that capital was also a constraint. For two years in a row, breeding stock had been sold to pay for medical care. But Sr. Gomez' credit was very good and his investment potential high. His income was used only to support himself and his wife, spending $41,500 per person, in addition to recent housing investments. Even if Antonio were paid out of farm earnings, there should have been sufficient income for savings and farm investment.

Sr. Gomez believed that so long as the price of cacao stayed above $20 per kilo, he would still produce but with fewer inputs. He felt that it was "better to be working than to put your money in the bank," and would not consider other investments outside agriculture. During the low-price years of the mid-1960s he had "just worked a bit harder and depended on subsistence production." Antonio clearly had a more "businesslike" attitude about future investments and diversification; he made clear that he would probably go to work elsewhere if the price dropped much at all.

Sr. Gomez, who was retiring, had strained relations with most of his offspring. If Antonio kept working with him, and they got on well, chances were that he would inherit the remaining undivided land, and the other children would receive a cash inheritance. Already two sons had been given use rights over one hectare of land each. If Antonio left, Sr. Gomez would be tempted simply to give each child a hectare and sell the rest. He considered retiring to town. It is notable that in the 1970s, nearly all of his net income went to consumption, rather than new investment.

Summary of Case Studies

The HLB families had quite substantial capital assets, largely because of the highly inflated price of good cacao land, although the value of buildings and other infrastructure was also substantial. According to the CCT, the average farm size was 14 hectares, and on average the HLB farm produced two to three times more cacao than did family-labor-based farms and more than eight times as much as the minifundia. Cacao was usually the principal activity. Fourteen percent of the farms in the SST sample included livestock production among the two most important activities.

Production patterns were similar among hired-labor-based farms. Per hectare imputed cash costs (valuing family labor at $120 per day) were nearly the same for all three farmers described here, despite their very different economic philosophies. Also, they compared roughly to the $10,647 per hectare cost estimated by the Cacao Plan as "traditional production without rehabilitation."

This was despite a relatively low level of chemical input use per hectare ($530-1,890 compared to the recommended $4,731). This low level was surprising given that these farmers had the greatest access to capital, and that the prices for chemicals were inflated. According to the CCT, the HLB farmers in 1978 tended to have more spraying and other types of cacao equipment and used more fertilizer than other farm types. More farms had their cacao and shade trees in good condition than the minifundia, though not as many as among family farms. The critical limiting factor appeared to be labor costs and labor availability for chemical application and cultivation care, and productive practices had deteriorated with rising wages.

Except for Sr. Prado, depreciation costs were not high relative to total costs. But costs for the use of equity capital, estimated at a very low seven percent (banks were then paying 15-25 percent on simple deposits) accounted for two-thirds to three-fourths of all costs for the three producers, not relating to family labor use. Including inputed costs of family labor, the figure was 40-50 percent, approximately that in the "recommended" budget.

None of the farmers had a major cash flow problem, largely due to diversification of farm enterprises. All three case study producers covered cash costs and depreciated costs and still showed cacao earnings. Sr. Ramirez and Sr. Gomez both covered equity cacao costs as well, earning $8.43 and $7.24 on every peso invested. Even Sr. Prado earned $2.50 per peso invested. Net family income was well over recorded levels of consumption for Sr. Prado and Sr. Gomez; Sr. Ramirez didn't quite break even, but he was supporting several households. Although all three were either reducing cacao production or anticipated doing so in the near future, the cacao enterprise was in fact extremely profitable for these farmers. To explain this contradiction, one must look at the key factors affecting farm management decisions for hired-labor-based producers.

HIRED-LABOR-BASED FARM FAMILY ECONOMIC STRATEGIES

The HLB farm families shared a number of common production, investment, and employment strategies and tended to have very specific responses to the economic changes brought about by the petroleum boom. Their basic goals were to maximize profits, minimize labor supervision, and to promote upward mobility in family employment.

Profit Maximization

The hired-labor-based producers sought profit maximization in a conscious way and were not excessively risk-averse. The farm was managed in such a way as to produce as much cacao as possible with the least inputs. In order to justify farmer interest in improved productivity, the cacao enterprise had to produce profits comparable to or greater than the average farm or nonfarm alternative, in terms of returns to capital invested and returns to family labor. Also, when returns to cacao were less than those in other farming or off-farm activities, the farmer responded by transferring resources to those activities. Return to family labor was a less important consideration than return to capital.

Mobility of Assets

These families had sufficient assets that they were not restricted from entering alternative occupations or investment for reasons of security, access to cash, or means of production. A notable feature of the group had been the transfer of resources out of cacao in recent years, in favor of cattle production or nonagricultural investment. Despite the profitability of cacao indicated above, these other activities were more attractive to farmers. The phenomenon was especially interesting, since several years ago the trend was for businessmen successful outside of agriculture to purchase cacao farms with their nonfarm incomes and profits.

Minimizing Labor Supervision

Another factor pushing these farmers away from cacao was the desire to minimize their use of hired labor. This was not due to unusually high costs. Labor earnings per man-day in cacao were greater than the typical wage paid for cacao work, and (the exception of Sr. Prado notwithstanding) it was usually much higher even than the highest quoted wage for harvest, drain-digging, or other higher-paid work. Rather, unwillingness to supervise labor, the costs involved in search time, and uncertainty about stability meant that labor would rarely be hired to the point where marginal returns equalled the wage. Indeed, given the multitude of factors which affected yields, the marginal value product of hired labor was impossible for farmers to estimate even in a rough way.

In the days when labor costs were quite low and fairly stable in Tabasco, before the 1970s, considerations of labor use in enterprise decision making were minimal. Labor demand was directly related to cacao prices and yearly harvest levels. This relation no longer held. Yet many of the larger employers were philosophically outraged at the evolution of high labor payments, which were felt to be undeserved or likely to be wasted. They were averse to dealing with large numbers of workers whose bargaining position was suddenly strong, while their own position had weakened.

Unlike other types of farm families in which total family income was in part derived from off-farm wages, increased labor prices were always negative for the hired-labor-based family. Higher general wages did improve the internal family bargaining position of sons who worked on the farm, however.

There was a clear inverse relation between access to capital and use of family labor. For example, cash costs per hectare on the Gomez farm were only half that of the Ramirez farm and a quarter that of the Prado farm. Family labor as a percentage of total labor was two-thirds, four-ninths, and one-twentieth, respectively. Were wages to decline sharply, hired labor would probably be used to substitute for family labor almost entirely.

Otherwise, family and hired labor were to a certain extent complementary. Even minimum maintenance on these larger plantations required use of hired labor. Increased producer family labor use, over a wide range of inputs, implied greater demand for hired labor--at least during the harvest period. Similarly, increased use of hired labor to increase production tended to increase the demand for work from family women and children for pod-breaking, washing, drying, and sorting activities.

Family Employment Strategy: Upward Mobility

The HLB farm households were less tightly bound to the farming way of life than were those campesinos for whom the land represented security in a very fundamental sense. HLB family employment strategy, even where families were dedicated to farming, included a continual search for improved personal and family options, as illustrated by the SST employment statistics.

On the typical HLB farm, the farmer and his family performed about half of the required farmwork. The exact proportion was determined by the demographic stage and related issues: number of older sons available; the sons' decisions to work on or off the farm; total labor demanded on the farm; and the degree of farmer participation, i.e., his choosing to work on the farm rather than in alternative businesses.

On smaller farms family women participated in cacao production, but this activity was minimal and rarely included physically demanding activity that might be seen as a sign of diminished family status. On the other hand, many educated daughters sought employment in services, commerce, or the professions that conferred considerable status.

Fifteen percent of heads of household on HLB farms did cacao farming as a secondary occupation. These heads of household tended to have a diverse employment background. Those who had not recently entered farming were jornaleros in their early years. But in 1980 one rarely saw a HLB farmer over the age of 30 working as a low-status jornalero.

A high proportion of older sons did perform agricultural and cacao wage labor, but this tended to be on the family farm, rather than on other properties. A tension existed between father and sons about the role of sons on the farm. After age 16 or so, sons became concerned with earning their own incomes and determining future professions, and we see the development of multiple individual strategies rather than a single "family strategy" of investment and employment. According to the SST, 18 percent of the total EAP from HLB farms--nearly a quarter of nonhead of household EAP--worked outside the agricultural sector in 1979.

Basically HLB farmers and their families considered their principal role on the farm, once reaching their mid- to late-20s, to be that of manager rather than worker. As a result, few family members *not* in a managerial role remained on the farm. Those HLB workers who left cacao farming usually changed to skilled trade occupations or employment in commerce and services.

Farmers who did very well financially encouraged their children to become upwardly mobile and to seek a future off the farm. As a result, they became totally dependent on hired labor. To some extent this reduced the pressure to subdivide properties. On the other hand, the greater independence of sons caused by the changing economic environment forced many families who needed and wanted children's participation on the farm to depend instead on hired labor.

HIRED-LABOR-BASED FARM HOUSEHOLD RESPONSE TO ECONOMIC CHANGE

The petroleum boom disrupted the cacao economy in ways that are likely to permanently reduce production by HLB farmers: the economic position of cacao relative to nonfarm activities in the state has declined; and the cost and availability of labor greatly constrain the potential for growth for farmers dependent on hired workers.

Hired-labor-based cacao producers seem to have been highly responsive to relative prices historically. In the past, if the long-run price of cacao relative to other potential products looked positive, the hired-labor-based producers continued basic maintenance operations, despite low prices. This seems to have been the case during the late 1970s, with those families having lots of family workers actually intensifying production somewhat. Otherwise, they were strongly tempted to replace plantations with pasture, given the less demanding labor requirements of cattle-raising, and there are large extensions in the Sierra of ex-cacao land now used for grazing. As cacao prices decline, one should expect to find marginal cacao land in the hands of HLB producers doing the same in the 1980s.

If wage rates relative to the cacao price should drop, hired-labor-based producers can be expected to increase their use of labor in cacao production. Although profits would improve, it is unlikely that there would be a corresponding improvement in productivity (i.e., use of more inputs) given the importance of labor costs in application of most new technologies. If cacao prices increase, labor use on the HLB farms would increase, but not at as fast a rate as the expected increase in nonlabor inputs.

AGRICULTURAL PROGRAMS FOR HLB CACAO FARMING

Technical assistance programs were helpful for hired-labor-based farmers, but the basic thrust of the Cacao Plan assumed unrealistic access to labor, as did the recommended farm budgets. The profit level needed to inspire hired-labor-based producers to add new inputs was greatly underestimated.

Innovative solutions to labor scarcity needed to be found --for example, traveling work units composed of landless and minifundia workers which charge normal wages but guarantee high quality work. The most serious potential problem--abandonment of plantations, or change to a more extensive use of land--might be handled by new land and cultivation rights arrangements.

The HLB family tendency to move many or most children into nonfarm activities dovetails nicely with long-term Tabasco State Plan employment objectives. However, the Plan did not sufficiently recognize the potential problem of a continued decline in cacao production by the HLB producers, who, after all, were still responsible for a large share of state production.

THE ROLE OF HLB PRODUCERS IN A PETROLEUM ECONOMY

The very source of the capitalist farmer's advantage in stable economic conditions--rapid and extensive response to price incentives--can be the source of poor performance in difficult times or under uncertain economic conditions, such as petroleum boom. A significant response to relative price changes in wages and in crops, and farm:urban price relationships can lead to large-scale declines in production and transfer of accumulated farm capital to urban investments. The internal rationality of HLB (and capitalist farming in general) continually pressures the farmer to reduce his wage burden. Producer attempts to change production technology or land use in response to a short-term, but sharp, rise in real wages, can lead to a long-term reduction in the local employment potential for crops where labor-saving technology is available.

Whether traditional campesino, or sophisticated capitalist, farmers who must depend upon hired labor for a large part of their production will be particularly vulnerable in an economic environment of high wages, high costs, and high off-farm investment opportunities.

5

Family-Labor-Based Farm Production Strategy

The Tabascan family-labor-based (FLB) farm producer represents the smallholder population with sufficient land to support a family. In 1978, they accounted for over a quarter of all cacao producers in the state and a third of all production--and that proportion was growing.

Family farm cacao production has probably been the predominant historical form of cacao production since the time of the Mayans. Given the relative abundance of land, there was no reason why producers should be unable to plant as much cacao as family labor resources would permit. The areas today where this type of farm predominates are zones newly opened up late in the last century, in areas of colonization, and where land was granted to ejidos. The most common story told in the regional survey was that parents or grandparents moved from an area of dense population in the Old Chontalpa to frontier areas where they cleared land, planted subsistence crops, and then established cacao plantations on all land that could be opened through use of family labor.

Because these areas were sparsely populated, there were few landless laborers for hire. Community mutual aid patterns between families, compadres, or neighbors developed to meet peak labor demand. The typical lack of schooling for children meant that, at a young age, they participated in cacao harvest and pod-breaking.

In the past couple of decades, with increasing population density, the rise of minifundia, and improved communication, interactions between the three types of producers have increased markedly, including the use of hired labor on the FLB farm and hiring out to largeholders when cash is short.

The FLB farm mode of cacao production was closest to the conceptual "peasant mode of production." The family is the basic unit of both consumption and production, and the family economy changes over time with the demographic cycle. For Tabascan cacaoteros, the paradigm had to be adjusted for active labor and product markets. Concern for subsistence and the secure reproduction of family and farm were paramount, but the last generation of farmers found these conditions satisfied through combined production for home use and cash sales of cacao. Hired labor was used, but it comprised only a small proportion of total labor.

Demographic and employment traits were tightly interwoven; Chart 5.1 summarizes the relevant data from the SST. The demographic cycle as it operated among Tabascan cacao farming families can be summarized as follows. Young men, sons of cacao farmers, still single or newly married, worked on their father's plantations but also as jornaleros to earn cash for future investment or setting up a household. Their sisters also helped on the farm and earned cash providing washing, sewing, or other services.

At marriage, the son received his own parcel of land to use. As soon as it provided sufficient income to support his family, he ceased jornaleando except as a source of emergency money.* He sometimes continued to help his own father in these early years. As soon as his own children were old enough to help in the fields, farm production went up markedly, and so did the need for and ability to afford hired labor.

When sons were in their late teens, the need for hired labor went down. The trade-off between sons working for the father or off-farm depended on relative earnings, desire for independence, and family ties. As children grew older and devoted more attention to their own affairs, the farmer again had to seek supplementary hired help. The SST shows a high incidence of migration for these families, particularly for individuals aged 16-20. If, over the years, the family acquired considerably more cacao area via inheritance or purchase, or the combining of assets, it could develop into a hired-labor-based farm.

The FLB farm household was typically headed by a middle-aged couple with several children out of elementary school, just breaking the severe cash constraint caused by a large number of dependents per family worker, blessed with many willing hands whose labor need not be paid. On average, the family farm had more working age members than other family types. Just over six percent of heads of household in the SST sample were female.

---

*The pattern had changed slightly from the days when most young men were establishing new plantations, rather than improving established ones; there was not the same need for supplementary income while waiting for trees to first produce.

Chart 5.1 Demographic and Employment Characteristics of Family-Labor-Based Cacao Farm Households

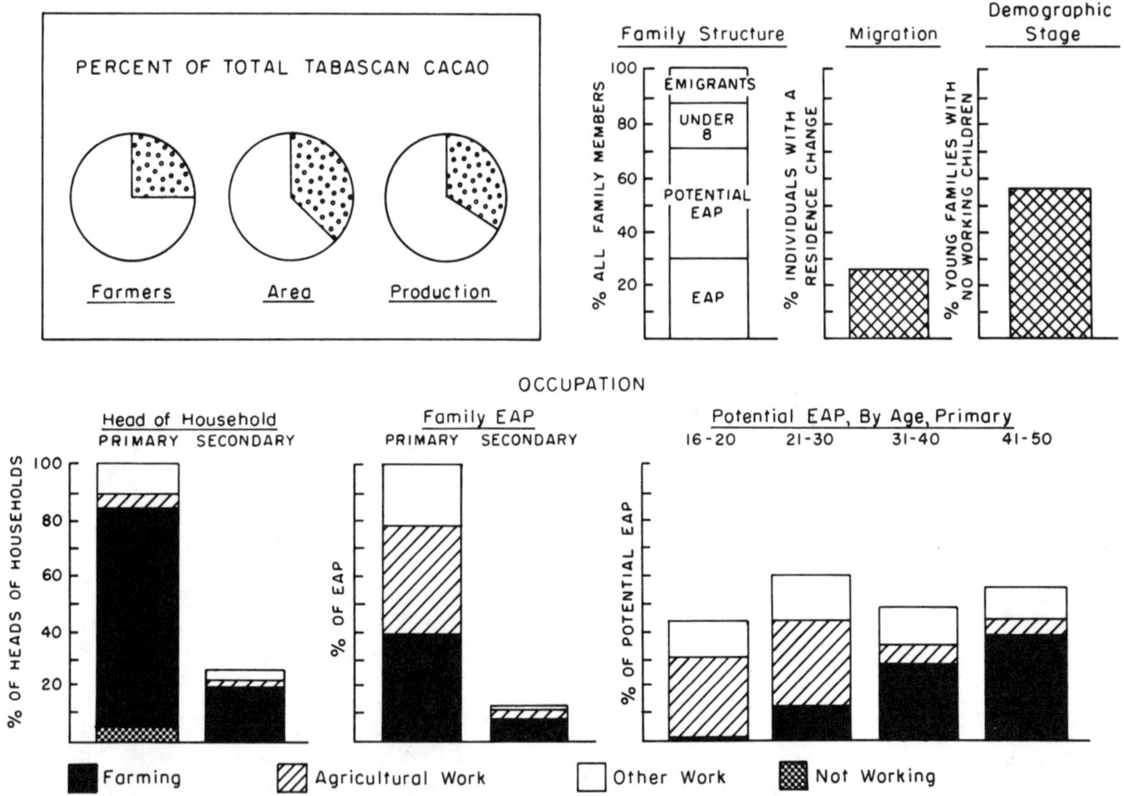

Source:

TYPICAL FLB FARM HOUSEHOLDS: CASE STUDIES

Three representative family-labor-based households are those of Sr. Rama, a private property owner newly turned ejidatario in Galeana II in Teapa, and Sr. Luna and Sr. Sandoval, ejidatarios in Huimango II in Cunduacán. Data collection for another family in Quintana Roo was not completed. Tables 5.1 through 5.3 summarize data on production, costs, and employment on these family cacao farms.

Sandoval Family

Sr. Sandoval, 50 years old, was born in Ranchería Huimango II. His parents received land (five and three-fourths hectares) on the ejido when it was formed to meet the needs of the expanding population of Huimango I. Sr. Sandoval inherited the land in 1969. The family planted maize initially, soon planting shade and cacao nursery trees. By 1960, 1,800 trees were planted; by 1970, 2,500; by 1979, 100 banana trees and forty pepper trees had been added on remaining vacant land. The family had eight head of cattle which were killed in the hoof and mouth disease epidemic in 1952.

Sra. Sandoval, 48, was born nearby. Three of their ten children were married (ages 20-24) and lived on the ejido. The two sons worked for PEMEX and helped their father on the farm. A nephew had lived with the family until a year earlier; he was currently employed in construction. Of the seven children residing at home, the eldest sons helped their father, the eldest daughter her mother. Three, aged 10-14 were in school; the two youngest were preschool. The older children had not attended school beyond the fourth grade.

Sr. Sandoval's cacao plantations comprised his entire assets, which were worth \$.5 million. Had he been a private farmer, the value of the land would have added another \$.5 million. The family also had a small number of small livestock. Of total gross income for the family in 1979-80, the farm provided three-fourths, almost all from cacao. One thousand two hundred dollars were earned from infrequent off farm jornalero work by sons Pascal and Guadalupe; \$16,200 was earned by the sons working for PEMEX. Table 5.4 shows household income statistics.

Hired labor performed 11 percent of all man-days of work on the farm and was responsible for 48 percent of cacao and 40 percent of all farm costs. Sr. Sandoval performed 26 percent of the work and his family the rest. The farm utilized one man-equivalent of labor during the months of May, July, and August; two in September and March; three to four during the rest of the year.

Sr. Sandoval faced two major problems during the 1979-80 seasons. His plantation had a severe cacao fungal infection,

Table 5.1 Inputs for FLB Cacao Farms, 1979-80

	Sandoval Farm	Luna Farm	Rama Farm	Recommended[a] w/o Rehab	Recommended[a] with Rehab
Productivity					
Number of trees	2,500 (435/h)	3,500 (500/h)	3,000 (500/h)	600/h	600/h
Age 1	40 yr - 1,400	35 yr - 1,000	15-20 yr - 2,000		
Age 2	20 yr - 500	25-28 yr - 2,200	10 yr - 500		
Age 3	16 yr - 600	2 yr - 300	6-7 yr - 500		
Pdn/tree	.68 - .77	.557 - .675	.346 kg	Average 1.3 kg[b]	
Pdn/Hectare	340 ($17,000)	278.6-308.7	173 ($8,650)	637	835 kg
				Average 538 kg[b]	
Man-Days/hectare	135 - 163	27.3	59.1	81.5	30
Pruning, clearing	28.8	4.5	23.0	21	24
Drains	1.8	.5	11	3	6
Fertilizer	7.1	--	10.6	2.5	10
Insecticide	.5	1.2	.6	12	20
Fungicide	28.8 57.6	1.1	1.3	15	30
Harvest & Pod-Breaking	68	20	12.6	28	40
Inputs/hectare					
Fertilizer					
Urea	--	2	--	--	2
18-9-18	--	--	1	--	--
15-15-23	--	--	--	--	--
Floren	2	2	--	--	1-2

Nutrisol	--	--	--	--	--
Gro-Green	--	--	1	--	--
Crescal	--	--	--	--	--
Insecticides					
Trident	--	--	--	--	--
Metilico	2	1	2	--	2-3
Foley	--	--	--	--	--
Folidol	--	--	2	--	--
Folimac	1	--	--	--	--
Polidor	--	1	--	--	--
Fungicide					
Lime & Sulfate	8	4-6	3	--	5-8
Cuprasol	--	--	2	--	--
Pruning	1 major frequent minor	1 major + 1 minorminor	1 major + frequent minor	--	2-3

Note: kg = kilograms; h = hectare.

[a]These data are from estimates in the SARH, *Plan de Cacao*, 1979.

[b]These data are from the SARH 1978 Cacao Census.

Source: Compiled by the author, Cacao Farm Budget Studies, 1979-81, Tabasco, Mexico.

Table 5.2 Costs of Cacao Production on FLB Cacao Farms, 1979-80

Inputs	Luna Farm	Sandoval Farm	Rama Farm	*Recommended*[a] *Plan* Grijalva[b]	w/o Rehab[c]	with Rehab[d]
Labor	$ 1,000	$12,550	$ 7,200	$ 9,400	$ 8,100	$13,900
Interest	--	--	12,000	--	--	--
Inputs	9,360	13,290	8,598	5,075	2,547	4,731
Miscellaneous	--	8,000 (rent, transport)	4,000 (equip. transport)	--	--	--
Fertilizer:	--	100	2,860	3,200	1,500	2,366
Urea	--	--	(960)	--	--	--
18-9-18	--	--	(1,480)	--	--	--
Floren	--	(100)	--	--	--	--
Gro-Green	--	--	(420)	--	--	--
Nutrisol	--	--	--	--	--	--
Insecticide:	4,800	540	1,600	675	1,047	2,365
Folimac	--	--	(880)	--	--	--
Tridente	--	(380)	--	--	--	--
Metilico	(7)	(160)	--	--	--	--
Methyl	--	--	--	--	--	--
BHC	--	--	--	--	--	--
Foley	--	--	(720)	--	--	--
Polidor	(4,800)	--	--	--	--	--
Folidol	--	--	--	--	--	--
Fungicide:	4,560	4,650	2,638	1,200	--	--
Lime	(960)	(150)	(138)	--	--	--
Sulfate	(3,600)	(4,500)	(2,500) est.	--	--	--
Cash Costs	10,360	25,840	30,198	14,475	10,647	18,631

Depreciation	1,000		1,000			
Full Costs	11,360	25,840	31,198	14,475	10,647	18,631
Equity Deducted[e]	50,076	36,127	44,033	14,000 (10,500)	14,000 (10,500)	14,000 (10,500)
Costs with Equity Deduction	61,436	61,967	75,231	28,475 (24,975)	24,647 (21,147)	32,631 (29,131)
Number of hectares	7	5	6	1	1	1
Total Cash Costs	1,480	5,168	4,616	--	10,647	18,631
Family Man-days/hectare	32.7	95	83	--	--	--
Imputed Cost Family Labor	,925	11,400	9,960	--	--	--
Cash for Inputs	1,337	2,658	2,058	--	2,547	4,731
With Depreciation[f]	5,547	16,568	15,159	--	10,647	18,631
With Equity Deducted[f]	12,701	23,793	22,497	--	24,647	32,631

[a]The original figures valued a man-day of labor at $75 (Grijalva data) and $83 (Cacao Plan data). Here we have adjusted this to $100, the minimal going rate found in Tabasco during my field study. At the more typical rate of $120, the estimates would have been substantially higher.

[b]SARH, Comisión de Grijalva, Dirección de Estudios y Proyectos, "Rendimientos y Costos de Producción de Cultivos Tropicales," August 1979, pp. 10-16.

[c]SARH, "Plan de Incremento de la Producción de Cacao en el Estado de Tabasco," January 1979, Villahermosa, Tabasco, Cuadro 20. This represents the supposed "typical" production system in Tabasco before rehabilitation programs.

[d]Same source as above. This represents the production costs after farmers have instituted rehabilitative measures on their plantations (see text) suggested by the Cacao Plan.

[e]A higher estimated value for land is shown in the first figure; a lower estimate in parentheses.

[f]These include imputed cost of family labor.

Table 5.3 All-Farm Production Costs on FLB Cacao Farms, 1979-80

Inputs	Luna Farm	Sandoval Farm	Rama Farm
Labor	$ 1,000	$12,550	$ 17,200
Cacao	--	--	7,200
Cattle	--	--	10,000
Maize	--	--	--
Cacao chemicals & miscellaneous	9,360	5,290	4,818
Vehicle[a]	--	--	--
Cattle Feed, etc.	14,000	5,250	--
Veterinary services & medicines	1,500	--	--
Equipment	--	--	3,000
Cattle purchases	--	--	--
Interest on loans	--	(47,500)	12,000
Rent Land, Equipment	--	1,000 sprayer	--
Transport & fuel	585	7,000	1,150
CASH COSTS	30,145	32,090 (78,590)	37,948

Depreciation	3,000	--	2,666
Cacao	--	--	--
Cattle	--	--	--
Maize, etc.	--	--	--
Presents of products	--	--	--
FULL COSTS	33,145	32,090	40,614
Equity capital costs	66,213	36,127	84,533
Cacao	50,076	--	44,033
Cattle	--	--	--
Maize, etc.	--	--	--
COSTS with deduction of 7% on equity capital	99,357	68,217	125,247

[a]This value includes tractors, trucks, etc.

Source: Author, Cacao Farm Budget Studies, 1979-81, Tabasco, Mexico.

Table 5.4 Family-Labor-Based Cacao Farm Household Gross Income, 1979-80

Source	Rama Family	Sandoval Family	Luna Family
Cacao	$ 61,480	$ 22,158[a]	$128,866
Poultry	--	15,000[b]	--
Garden	10,000	--	--
Cattle	--	--	8,800
Rent pasture	11,520	--	--
Pig sales	--	3,000	--
Milk	--	--	--
Pepper	--	540	--
Bananas	--	2,400	--
Miscellaneous	--	138	--
TOTAL Farm Cash Income	$ 83,000	$ 43,236	$136,866
Family produce consumed	1,300	8,400	2,040
Change in livestock inventory	--	--	12,700
Miscellaneous	--	47,500[c]	--
TOTAL Farm Labor Earnings	$ 84,300	$ 99,136	$151,606
Off-farm agricultural	--	1,200	--
Nonagricultural	96,000[d]	--	--
Loans	--	--	--
TOTAL Income	$183,300	$116,506	$151,606

[a]Direct sale.

[b]Sale to creditor.

[c]Additional worth of cacao sold to creditor (see text).

[d]Net earnings from store.

Source: Author, Cacao Farm Budget Studies, 1979-81, Tabasco, Mexico.

Table 5.5A Family and Farm Income of FLB Cacao Farm Households, 1979-80: Sandoval Family

Income Measure[a]	Cash Income[b]	Results: Labor Earnings[c] without Equity Return Deduction	Results: Labor Earnings[d] with Equity Return Deduction
a. Net farm ___	$12,146	$20,546 (60,046)	$ -15,631 (31,869)
b. Net off-farm agricultural	1,200	1,200	1,200
c. Net nonagricultural ___	16,200	16,200	16,200
d. Net total[e] ___	29,546	37,946 (85,446)	-31,000 (49,269)
e. Net farm ___ per hectare	2,112	3,573	-2,718 (11,834)
f. Net farm ___ per man-day on farm[f]	17	28 (93)	-25 (43)
g. Net farm ___ per family man-day on farm[g]	19	33 (108)	-25 (51)
h. Net nonagricultural ___ per family man-day worked outside agriculture[g]	120	120	120
i. Net agricultural off-farm ___ per family man-day worked[g]	180	180	180
j. Net total ___ per family man-day worked[g]	41	53 (119)	0 (69)
k. Net cacao ___ per man-day in cacao[f]	17	17 (85)	-34 (34)
l. Net cattle ___ per man-day in cattle[f]	--	--	--
m. Net pig ___ per man-day in pig work[f]	--	--	--

Table 5.5A Continued

Income Measure[a]	Cash Income[b]	*Results* Labor Earnings[c] without Equity Return Deduction	Labor Earnings[d] with Equity Return Deduction
n. Net __ ___ per man-day in __[f]	--	--	--
o. Net cacao ___ per family man-day in cacao[g]	20	20 (99)	-39 (40)
p. Net cattle __ per family man-day in cattle[g]	--	--	--
q. Net pig __ per family man-day in pigs[g]	--	--	--
r. Net __ ___ per family man-day in __[g]	--	--	--
s. Percent labor in cacao expenses[h/j]	49.2	49.2 (49.2)	20.7 (20.7)
t. Percent labor in whole farm expenses[h/j]	40.4	40.4	18.9

[a]The blank spaces below refer to the specific income measure used--cash and labor earnings, with and without equity deduction; the figures are found in the appropriate column to the right.

[b]Cash income refers to costs and returns to the family only in cash.

[c]"Labor Earnings" (without equity deduction) is a measure of income that includes several noncash components such as : inventory changes, family use of farm products, etc.

[d]"Labor Earnings" with equity deduction is similar to the simple labor earnings measure, but it also adds the cost of tying up the family's capital in the farm enterprise. It enters, as a cost, the interest that could be paid to the family by a bank were all assets to be sold for cash and deposited, without the risks entailed in a farm enterprise. The very low rate of 7 percent is used for our calculations.

[e]Total Income of whatever measure is the sum of farm income, off-farm agricultural income, and off-farm nonagricultural income.

[f]Income per man-day is a rough measure of average labor productivity including both hired and family labor use. To compute this, the wage bill is subtracted from total costs.

[g]This is the average value of a day of labor performed by family members. For this purpose, the wage bill *is* included in costs.

[h]The value in parentheses represents the higher of two cost estimates for cacao production.

[i]This pig budget is probably inaccurate. If we assume net returns at $1 million (based on unreported costs) and a doubling of the very low reported labor input to 800 man-days per year, then returns per man-day are reduced to $1,250.

[j]This does not include the imputed value of family labor.

[k]The value in parentheses represents costs (or returns) in the form of a changed animal inventory.

Source: Author, Cacao Farm Budget Analyses, 1979-80, Tabasco, Mexico. Note that the costs, returns, and labor use summarized here can be found in preceding chapter tables.

but this was treated at negligible cost by a Cacao Plan/Cacao Association special spraying squad. Worse, illness forced the family to borrow a large sum of money from a local merchant with the condition that they sell a certain amount of the cacao harvest at an extremely low price. Total consumption expenditures for the Sandoval household in 1979-80 were over $100,000, over a third of which was the emergency medical expense. The rest went to support the family at $7,333 per person.

Under these conditions, only 41 percent of income came from the farm; labor earnings with equity return were negative; labor earnings per family man-day worked in cacao were only $20 ($17 for hired labor). Interest on the borrowed $15,000 was 300 percent. By contrast, valuing all cacao produced at market figures, 80 percent of family income came from the farm; the value of labor earnings per man-day in cacao was about $85; of a family man-day, $40.

The main production constraint for the Sandoval family was cash. Their technology seemed to be good, and Sr. Sandoval consulted with all the scientific people in Tabasco, listened faithfully to the Cacao Plan's radio program, and attended a special cacao production class. He had received bank credit for five years until 1978-79, when a bumper harvest made it unnecessary. He planned to seek credit again in 1980 to cover fertilizer and hired labor expense. He needed to improve drain control, increase weeding, begin soil fertilization (which he badly wanted to do), and improve labor efficiency in fumigation and harvesting.

In the medium-term, his sons could provide more labor, but the elder may decide to devote more time to PEMEX activities. A severe labor problem could develop unless José remained full-time. Yet José in 1979-80 brought in more cash income ($180/day) from his off-farm job than he "produced" in farm work. Total net cash family income was $30,000 or US$1,300 (under $500 after devaluation).

Luna Family

Sr. Luna, 76, was born in Jalpa town. He lived with his wife Elena and two sons and two daughters aged 17-26. Three married sons, aged 35-43 worked with their father on the land; the fourth son had a small plot of his own. The three married sisters who lived nearby were housewives. All the children were born in Huimango II; none had more than a third grade education.

Initially landless, Sr. Luna was one of the original settlers on the ejido in 1939, receiving rights to six hectares in 1943. During the Trujillo administration he received help from the Ejidal Bank to plant cacao. Maize for subsistence was produced until the mid-1960s, rubber and coffee until the late 1970s. The three married sons who helped Sr. Luna received land in Huimango I ten years ago, but it was of low quality, unsuitable for cacao.

Total capital on the Luna farm was valued at nearly $1 million pesos (US$42,000), mostly in the value of the cacao trees. The ejidal land did not "belong" to the family, therefore only improvements counted as assets. If land value were included, his total assets would increase to $1.2-2.0 million pesos. The small herd of livestock was worth $91,000. Total cacao production during 1979-80 was 2,161 kilos of dried beans.

Sr. Luna himself worked less than six percent of all man-days worked on the farm; his sons worked nearly all the rest. The percentage of hired labor in farm production for Sr. Luna was only two percent. This expenditure accounted for ten percent of cacao cash costs. Most hired workers were related to the family and lived on the ejido. Farm labor needs were unevenly distributed: May-October, the farm utilized only half a man-month equivalent; November-April, one and a half to two and a half man-month equivalents. The 24 hectares jointly operated by the three men thus seemed to be underutilized. Sr. Luna was semiretired, active mainly in farm administration.

Cacao yields in 1979 were lower than normal, because the farm had suffered a severe insect infestation. The family had probably fertilized, at most once (Mr. Luna could not remember), and the time spent in general spraying was much less than recommended. Cash seemed to be the most important constraint to increasing production.

Assets were tied up in cattle and land, and the sons were unwilling to pay more attention to cacao--causing a labor shortage either for lack of inputs, or because they had some alternative undisclosed employment. Cacao played only a minor role in individual family member strategies. To reach recommended levels, cash spent on inputs would have had to increase fivefold.

Family cacao farm income data are shown in Tables 5.4 (p. 160) and 5.5B (p. 166). Sr. Luna and his sons sold their cacao jointly, and he distributed one tenth of the proceeds to each son, retaining the rest for himself. Returns per hectare were $6,650, giving a value per family man-day worked of $136-174. This value jumped to $150-192 when simple labor earnings were computed; labor earnings with equity return deducted reduced the value of the family man-day worked to only $29-71. This did not include potential income from selling steers in the near future, although half of cash costs were going to the livestock operation. Total net cash income amounted to $114,000, or US$5,130 (US$1-2,000 after devaluation).

Table 5.5B Family and Farm Income of FLB Cacao Farm Households, 1979-80: Luna Family*

Income Measure[a]	Cash Income[b]	Results: Labor Earnings[c] without Equity Return Deduction	Results: Labor Earnings[d] with Equity Return Deduction
a. Net farm __	$114,911	$126,651	$60,438
b. Net off-farm agricultural __	--	--	--
c. Net nonagricultural __	--	--	--
d. Net total[e] __	114,911	126,651	60,438
e. Net farm __ per hectare	6,550	7,036	3,357
f. Net farm __ per man-day on farm[f]	174	190	91
g. Net farm __ per family man-day on farm[g]	174	192	92
h. Net nonagricultural __ per family man-day worked outside agriculture[g]	--	--	--
i. Net agricultural off-farm __ per family man-day worked[g]	--	--	--
j. Net total __ per family man-day worked[g]	175	192	92
k. Net cacao __ per man-day in cacao[f]	208	205	117
l. Net cattle __ per man-day in cattle[f]	--	--	--
m. Net pig __ per man-day in pig work[f/i]	--	--	--
n. Net __ __ per man-day in __[f]	--	--	--
o. Net cacao __ per family man-day in cacao[g]	210	207	118
p. Net cattle __ per family man-day in cattle[g]	--	--	--
q. Net pig __ per family man-day in pigs[g]	--	--	--
r. Net __ __ per family man-day in __[g]	--	--	--
s. Percent labor in cacao expenses[h/j]	8.7	7.7	1.6
t. Percent labor in whole farm expenses[h/j]	3.3	6.9	1.2

*For *Notes* and *Source*, see Table 5.5A, pp. 162-63.

Rama Family

Sr. Rama, 57, and Sra. Rama, 51, were native Teapanecos, originally from town, who worked full-time in their dry goods store until 1976. At that time they bought three hectares of land on a ranchería, which they sold in 1978 to buy six hectares of run-down cacao land and eight hectares of pasture on the ejido. They lived on the farm with Sra. Rama's mother (79) and two sons, Sergio (27)--a full partner with Sr. Rama in the farm--and Javier (13). Four other children, 21-29, were scattered throughout the state and northern Mexico. Two were studying agronomy; one was an accountant; another was an architectural assistant. Sra. Rama and Sergio worked full-time on the farm.

Family assets in land improvements (mainly cacao trees) were worth $670,000 pesos in 1980. The house and other constructions were worth another $181,000. With equipment, total assets were worth US$38,655. Were the property private, the land assets alone would have been worth between $875,000 and $1.29 million. During 1979-80 1,040 kilos of dried cacao beans were produced.

Of the family's gross income only 11 percent was from cacao and only 15 percent from the farm. The rest still came from the family store in Teapa. Total net cash income was $138,000, or US$6,200 (less than US$2,000 after devaluation). During 1979-80, the Rama family spent almost $64,000 for consumption. Of this a third went to support children in school. The rest was for the family of five resident in Galeana, for an average expenditure of $8,394 per person. Total consumption was actually higher than this, because Sr. Rama had access to wholesale goods at his store, which kept total food expenditures extremely low.

Sr. Rama hired 16 percent of farm labor, which represented a third of cacao costs. This was primarily for nonproduction operations, construction, machine work. None of the hired laborers was a relative. Sergio was the principal worker, although Sra. Rama performed a very large share of the work, including much heavy labor, and Sr. Rama worked when he could.

Two workers were needed on the farm in all months except July, August, and September (when one was required) and February, March, and April (when four to five were required). Sr. Rama was paying a great deal of attention to basic investments in storage and drain construction, which took time away from cultivation. But given that, he did well and was obviously interested in reaching "recommended" levels of input use as soon as time and complete plantation rehabilitation would permit. He already used both soil and foliar fertilizer.

Of net total cash income for the Rama family (Table 5.5C), a third was farm income from cacao. The value of a

Table 5.5C Family and Farm Income of FLB Cacao Farm Households, 1979-80: Rama Family*

Income Measure[a]	Cash Income[b]	*Results* Labor Earnings[c] without Equity Return Deduction	Labor Earnings[d] with Equity Return Deduction
a. Net farm __	$ 42,552	$ 41,186	$ -43,447
b. Net off-farm agricultural __	--	--	--
c. Net nonagricultural __	96,000	96,000	96,000
d. Net total __[e]	138,552	137,186	52,553
e. Net farm __ per hectare	2,836	2,745	-2,896
f. Net farm __ per man-day on farm[f]	71	69	-73
g. Net farm __ per family man-day on farm[g]	85	82	-87
h. Net nonagricultural __ per family man-day worked outside agriculture[g]	333	333	333
i. Net agricultural off-farm __ per family man-day worked[g]	--	--	--
j. Net total __ per family man-day worked[g]	156	155	59
k. Net cacao __ per man-day in cacao[f]	68	66	-32
l. Net cattle __ per man-day in cattle[f]	--	--	--
m. Net pig __ per man-day in pig work[f]	--	--	--
n. Net __ ___ per man-day in __[f]	--	--	--
o. Net cacao __ per family man-day in cacao[g]	82	79	-39
p. Net cattle __ per family man-day in cattle[g]	--	--	--
q. Net pig __ per family man-day in pigs[g]	--	--	--
r. Net __ ___ per family man-day in __[g]	--	--	--
s. Percent labor in cacao expenses[h/j]	29.7	28.8	12.1
t. Percent labor in whole farm expenses[h/j]	42.5	39.9	13.5

*For *Notes* and *Source*, see Table 5.5A, pp. 162-63.

man-day of work on the farm was only $75, which rose to $90 per family man-day--still hardly competitive with agricultural labor rates in the area. Even if Sra. Rama's labor were assumed to be only half as productive as the son's, the value of the family man-day rose to only $108. By contrast, Sr. Rama's work in the store was worth $333 per day.

If we deduct the long-term investment work (a third of all cacao labor for 1979), the value per man-day rises to $103 and per family man-day to $106. Labor earnings including deduction for return to equity showed a negative return, although only slightly so for cacao. In order for the value of a family man-day to rise to $100 per day, cacao yields would have had to double, which was quite feasible in the near future.

The Rama farm's main constraint was labor--hired labor was not yet profitable, yet family labor had to be directed to renovating the farm. Because there was little cash flow yet from the farm, Sr. Rama planned to remain with the store for a few more years; after that his labor would be fully available. The main jobs that needed to be done more intensively were pruning, shade clearing, and spraying--all of which had to take place during the family's busiest months.

Knowledge was possibly also a constraint, since Sr. Rama and Sergio had limited experience with cacao. They were eager to be educated and had the advantage of two family mambers who were agronomists. The cash constraint--caused primarily by building their new house--would be relieved shortly, and presumably more hired labor and more inputs would be purchased.

Summary of Case Studies

As a result of the availability of family labor, FLB farm households showed more intensive use of purchased inputs and of cacao cultivation than did the hired-labor-based farmers with their high labor costs. On our three case study farms, the relatively low yields were a result of infestation, and lower than desired quantities of chemicals due to cash constraints. Imputing a value of $120 per family man-day, expenditures were midway between the Plan's "traditional" and "recommended" levels. Despite these constraints, all had shown a pattern of rising cacao production and input use over the prior three-year period.

Cacao was the pivot of farm production for these households, but economic diversification was fundamental. For all our case study households, for example, whole farm costs were far above cacao costs. Cacao income helped to subsidize Sr. Luna's growing livestock enterprise, Sr. Sandoval's pig feeding, and Sr. Rama's new farm infrastructure.

Depreciation costs (the costs of replacing buildings and equipment) were negligible on these farms, but interest on equity represented nearly half the total costs for Sr. Sandoval and Sr. Rama, despite their being ejidatarios. None earned enough income to cover equity costs for production, although labor earnings were comparable with market wages for family labor. But this was not a source of concern to the families, who did not base their economic strategy on returns to capital. The wage paid to hired workers was generally higher than their average product, which implies that workers were employed only to break bottlenecks in family labor. Despite high income insecurity, difficult access to cash for inputs, and active family participation in the nonagricultural economy, family-labor-based farm families were increasing their commitment to cacao and attempting to intensify production. This can be explained by the economic strategies followed by families with their specific land and labor assets.

FAMILY ECONOMIC STRATEGIES

The case studies illustrate the basic goals of family-labor-based farmers: meeting family subsistence requirements; ensuring perpetuation of the farm between generations; maximizing inheritance to children; cooperating among the extended family; reducing risk; and maximizing returns to family labor.

A family production strategy to meet these goals can be hypothesized wherein on-farm agricultural and nonagricultural production, off-farm wages, and other activities of all members are coordinated to maximize the "implicit wage," family labor earnings. This goal is constrained by some consideration of risk and minimal demands for farm reproduction.

This is quite distinct from an objective strategy to maximize "profits," or return to investment or capital. It is also distinct from the Chayanov model of fixed cultural aspirations and isolation from land and labor markets. The greater economic security of having a crop much more remunerative than subsistence crops, and plentiful off-farm employment for meeting other cash needs, had temporarily relegated subsistence farm production to a sideline.

Living Standards

FLB cacao farm households had unprecedentedly high incomes in the late 1970s compared to the very modest, near-subsistence conditions obtained before. Expectations of material standard of living had risen tremendously. The most common use of cacao profits for this group of farmers was for brick house construction. Embarrassed at still living in a thatched home, Sr. Sandoval apologized: "We're not in the sticks anymore; Huimango II is getting citified; we ought to have a more civilized house." Income changes had led not to incremental change, but to transformation: the first use of home appliances, the first generation of children educated, and the first bus and transport service.

Was this new consumerism a drain on potential farm investment, or a stimulus to increasing farm productivity? Family cash incomes were still very low; for our three families they were between US$2,000 and US$3,500 in 1980 dollars. This was not very far above minimum subsistence requirements. If cacao prices should plunge, they may be expected to actively seek other cash-producing alternatives such as off-farm employment and to place more emphasis on the subsistence role of the farm.

Maximizing Returns to Family Labor

The family-labor-based cacao farm required one to two persons working year-round and three to five during peak times of the year. Where those extras could be gathered from a large family (in the middle and later stages of the demographic cycle), resident children were unpaid; older children shared in income. Non-resident children were usually paid if they provided a large share of the work. But for short periods--especially at harvest and pod-breaking time--work was more of a family ritual, and family members were not paid.

Young sons were rarely available any more for full-time farmwork. They were in school or sometimes in petroleum work, but they still usually participated in the cacao harvest. Few family members continued to work for largeholder cacao producers, preferring the higher return for work on the family plot. Heads of FLB farms practically never worked as jornaleros. If no sons were available to help, the farmer would "adopt" a young man as junior partner, who would later inherit part of the land.

Women participated in pruning and weeding, as well as in harvest and pod-breaking, but not in heavy work like spraying, ditch-digging, or tree-felling. In the early years of marriage they sometimes provided a large share of total labor. They were also generally responsible for care of small livestock.

The most striking aspect of family labor use for these farm households was its fluidity. The supply of family labor was extremely elastic. Sons held a wide variety of jobs on and off-farm. On the Huimango ejido, there were many families whose young men had travelled to the U.S., Yucatán, and northern Mexico in search of jobs, as well as other parts of

Tabasco. Many worked for PEMEX. New small-scale enterprises, such as cattle-raising, could be undertaken, dependent only on labor available on the weekends or late afternoons, when children were home from school or work. At critical points in the farming calendar the FLB farm had priority for family labor. At other times family members chose jobs that provided the greatest return on their work time.

Under such conditions, the warmth and loyalty in the parent-offspring relations determined the availability of family help. Due to the transformation of the Tabascan economy, sons could no longer be forced to obey their fathers, as they could in the days when farm inheritance was the most promising--or only--future. In 1980, fathers (at least the most successful farmers) paid their sons generously, supported them in seeking outside training or work, and actively wooed them for partnerships.

Statistical Employment Trends. These trends are seen quite clearly in the SST employment statistics. Of all cacao farm heads of household, those from FLB farms had the lowest rate of participation in the agricultural sector as a principal occupation, though they were nearly always self-employed.

Relative to other cacao families, this group had the lowest proportion of jornaleros and high proportions of paid and unpaid family workers; the lowest over-all participation in commerce and the skilled trades. Ninety-five percent of cacao workers were aged 9-30; between ages 20-40, FLB farm household workers included a very high proportion of skilled tradesmen. Activity rates in PEMEX were similar to those of minifundistas and nonfarm workers. Female participation in the work force was over ten percent, mainly in agricultural and unskilled work.

One third of all the EAP (a very high proportion) had a secondary occupation, predominantly in cacao farming for heads of household working in other sectors. Only a fifth of secondary jobs were for agricultural wage work. There had been a slight increase over time in the proportion of cacao landowners doing cacao farming as a principal occupation, as would be expected. But there was a steady increase in the percentage of all individuals working outside agriculture (especially in skilled trades and the tertiary sector), from 2.5 percent in 1965 to 9.8 percent in 1979.

Farm Inheritance

The sensitivity of parent-child ties carried important implications for farm continuity and inheritance. Traditionally, land rights passed from father to son at the father's death. During the 1970s, fathers who wanted their sons to be farmers had to respect their new independence and often offered land rights much sooner.

But land succession had long been associated with family feuds, occasionally resulting in violence--a fear commonly expressed by parents. Only rarely was a single child interested in cacao farming, with siblings uninterested and the farm succession clear. Given the predominance of assets in land, there was little else left to divide among siblings who forfeited the land. Even on ejidos, where legally the land could not be broken up, the "heir" usually just represented the flock who may or may not have worked the parcel as a single unit.

This subdivision threatens to turn the mass of Tabascan cacao farms into minifundia, which the next chapter will show are less likely to provide stable, high-input cacao production, than are FLB farms. A significant advantage of the new high cacao prices for these families has been the opportunity for parents to offer some sons an alternative inheritance to farming--an education, an apprenticeship, a help in business --and thereby prevent subdivision. Continuation of a healthy off-farm labor market appears to be critical to maintenance of economically viable cacao production units.

Managing Risk

The insecurity in cacao farming, of depending on a volatile international trade good when one's family is not far from subsistence, was high and had been made higher by the petroleum boom. Insecurities mentioned in Chapter 2 (pollution, expropriation) made investment look less promising and agriculture look less profitable in general. Community support networks had been weakened, and most farmers claimed that only their family, and occasionally the Cacao Association or some local businessmen, could be depended upon to help in a family emergency or sudden economic disaster.

The family response has been: to renew family maize production and production of other goods for home consumption; to transfer some resources from cacao to cattle or other small scale enterprises; to educate children for nonagricultural careers. Unlike hired-labor-based farm households, however, the FLB farm households continued to rely very heavily on cacao to meet cash needs. In the SST sample, although 78 percent of FLB farm producers had cacao as the principal product (only six percent had maize), three quarters of the farms had some secondary product, generally maize or livestock for fattening.

The farmholding was itself the most important source of long-term family security; it was guarded jealously. These families had a heavy stake in improving incomes and income security by increasing cacao yields, more so than any other group of farm households.

FAMILY-LABOR-BASED FARM RESPONSE TO ECONOMIC CHANGE

FLB farm households responded to changes in cacao price, input prices, and wages differently than did hired-labor-based producers.

The most severe constraint to production for these farmers was lack of cash, and hence their ability to purchase chemical inputs was restricted. Input subsidies on insecticide, fungicide, and fertilizer could be expected to induce sharp increases in their use. Because they were "chemical-hungry," their level of input application had more to do with income received in the prior year (i.e., availability of cash) than with the current price of cacao. The relatively low cost of marginal inputs of family labor for input applications meant that these could generally be more intensively used on FLB farms than on hired-labor-based farms.

If the price of cacao should rise, farmers would, for the reasons explained above, use more inputs, and hence increase production significantly. Should the price decline, farmers would restrict input use, and after a point, start looking for supplementary off-farm employment. Indeed, during the period of low cacao prices in the 1960s, almost all FLB farm workers earned a large part of their income off-farm. Their supply response to declining prices was slower than hired-labor-based producers.

The response to changing wages differed between those families with temporarily little available family labor and those with substantial family labor resources. Both types differed from hired-labor-based producers in that most increases in production come through a high ratio of labor to purchased inputs. The short-handed household had a fairly straightforward response. If wages declined (relative to the value of family labor on the farm), they hired much more labor per unit land than hired-labor-based producers under similar conditions, because they did not have the latter's problems with labor management. When wages went up, family finances suffered at harvest time. But the lack of a major alternative income meant that so long as cash or credit was available, these producers simply paid the going rate without reducing their hiring. Investments in drains or other infrastructure projects, if carried out by hired labor, did decline with a wage increase.

Families with many able hands had a much more convoluted response to wage changes, because they both hired in and hired out labor. When off-farm wages declined (relative to labor earnings in cacao), sons were willing to work more on the farm, and farmers correspondingly needed to hire less labor, and were even less likely themselves to seek short-term wage labor. During the 1970s heads of FLB farms rarely worked off-farm in agriculture, which contributed sharply to labor scarcity for largeholders.

When wages increased (relative to the value of family labor on the farm), sons were induced to work more off-farm. Total family income might not change, and basic family labor resources would still provide for farm production at lower levels, but the composition would differ. At very high wage levels, of course, producer family labor on the farm would diminish to mere maintenance operations, then increase slowly as wage rates fell. Once rates approximated the level of return to family labor for cacao, family participation on the farm would increase sharply--it is preferable to work for oneself rather than as a jornalero. As wage rates fell further, off-farm work would drop off until diminishing returns set in on the farm.

AGRICULTURAL PROGRAMS FOR FLB CACAO FARMING

A modest program of technical assistance is potentially quite appropriate for this type of producer, who has a fair amount of resources to put into cacao, plenty of labor, and lots of interest in cacao. The Tabasco policy of price stability and technical education has met an enthusiastic response in terms of production and new investment, despite threatened tenure security in petroleum zones. The central problems not satisfactorily dealt with in the 1970s Cacao Plan were the cash constraint and the threat of minifundization.

There was little official appreciation of the cost effects on a family-based operation of a jump in use of hired labor for increased input applications and harvest. Family strategy to deal with this may be quite diverse--for example, an increase in off-farm employment in order to come up with the needed cash. Provision of small amounts of short-term consumption credit to cacao farmers needed to be encouraged. This would get them over the troughs in cash flow and allow greater input applications even in cash-short months. The production credit supplied officially in 1979-80 was of such a large amount relative to family cash expenditures and incomes that many interested producers had to refuse it. Technology development oriented to these producers needs to be largely labor-enhancing, not labor-replacing.

The legal changes in land ownership patterns for Tabasco, just introduced in 1980, were intended to combat minifundization in the state. But they were designed as though for maize farming in the Central Plateau, not the tree crop farmland of Tabasco. Projected minimum plot size was too large. The five-hectare cacao farm was a highly viable production unit; in light of recent experience, it was far more stable than a 20-hectare farm. The ejidatario with six hectares of cacao must be legally allowed to employ workers on his farm. Accepted systems to reduce the number of land managers, and provide alternatives to family members who are bypassed by inheritance, were needed.

In cacao production and other intensive crop production, the family-labor-based farm--big enough for diversified production, small enough to induce intensive crop care, snugly intermeshed in the state's labor market as an important source of both workers and employment--seemed to be the best hope for a lasting Tabascan agriculture in the current and future petroleum economy.

THE ROLE OF FAMILY-LABOR-BASED FARMS IN A PETROLEUM ECONOMY

The mixed subsistence/commercial, family-labor-based farm sector can play a stabilizing factor in agricultural economies distorted and disrupted by a petroleum boom. Such households may lack sufficient resources to provide sharp increases in production when prices increase, but they also refrain from major farm disinvestment under most circumstances, including very low relative prices. During periods of economic upheaval, access to subsistence production minimizes their vulnerability, as does their ability to take advantage of off-farm cash strategies without significantly reducing on-farm production.

In Tabasco, FLB farm employment characteristics were the key to economic stability. The multiple employment and high temporary migration levels of the Tabascan FLB farm households did not seem to signal deterioration. On the contrary, they were a very effective adaptation to new economic conditions. They often used off-farm income as a source of capital for on-farm investment, and they showed a significant capacity for technical change and productivity increase in agriculture.

6

Minifundia Farm Production Strategy

The existence of large numbers of farms with too few cacao trees to support a farm family is a relatively recent phenomenon in Tabasco. It probably dated from the 1940s population increase and ejidal settlement, after which land hunger could not be satisfied through cheap land purchase or colonization. Instead, land was subdivided within families to provide for the needs of the younger generation. This occurred earliest in Comalcalco and Paraíso, the most densely populated cacao zones, but spread throughout the cacao-producing area of Tabasco. Because of legal impediments to subdivision on ejidos, 84 percent of minifundia farms were private properties. The cacao minifundia of Tabasco in 1978 represented nearly half of the state's cacao farmers, and their numbers were growing steadily, mostly at the expense of the family farm sector. In 1978, they accounted for about 30 percent of cacao production and were stable producers during the 1970s.

Chart 6.1 summarizes the demographic attributes of minifundia cacao households in Tabasco. Most families had 4-6 members, as did family farm households, but they tended to be much younger, with predominantly school-age children. Only a quarter of the families were composed of young married couples with none or a few young children. There were more families with 3-4 working age members (hence a low average potential dependency ratio), but where economically possible, children were in school.

The demographic cycle in minifundia farming was slightly different than for the family farm. The young single person travelled constantly looking for work. Once married, there was a period of very intense work as a jornalero or doing piecework type jobs with high focus on cash availability and, if possible, land was rented for maize production. After

Chart 6.1 Demographic and Employment Characteristics of Minifundia Cacao Farm Households

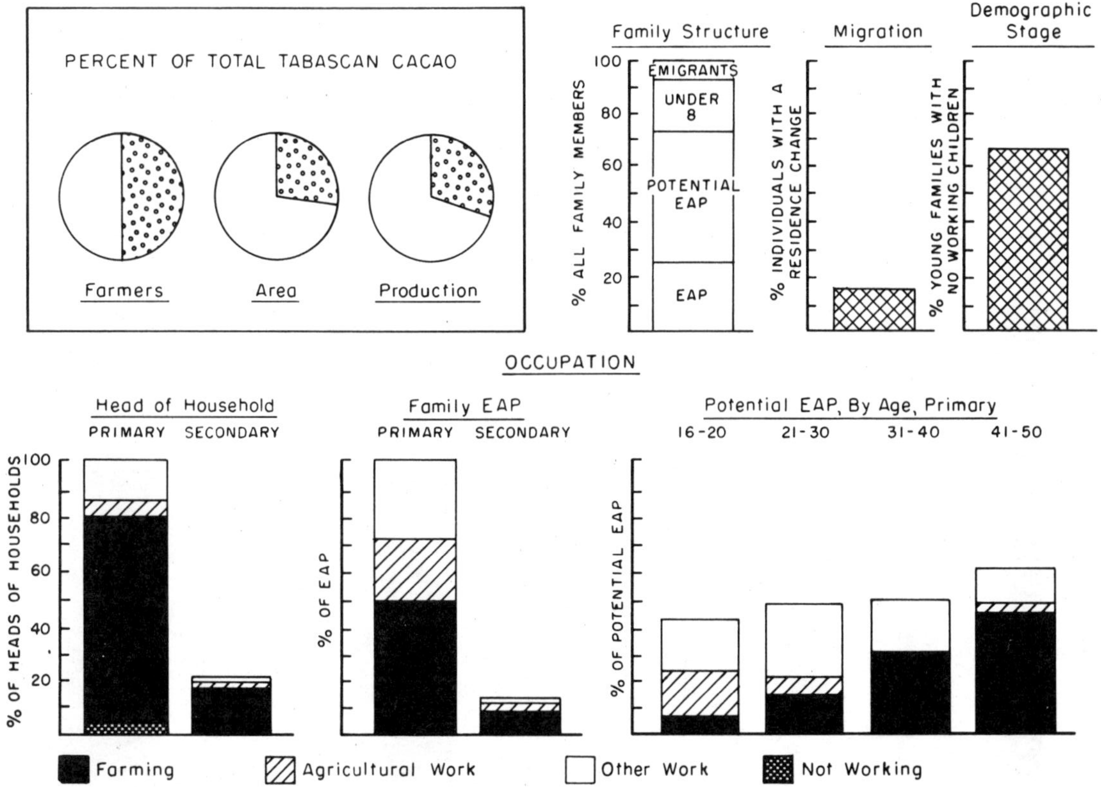

Source:

being granted the use or inheritance of cacao land, more time was spent on the farm, with increasing employment of children as soon as they reached productive age. Once children could largely support themselves through jornalero work, the parents turned full attention to the farm and retired from the off-farm labor market.

The long period when farming had to be sacrificed to cash income increased intergenerational family ties. Sons were more dependent on working with their fathers. Older sons could experiment with or prepare for different careers, falling back on the family farm in ebb work times and emergencies.

When land was more widely available in the state, the minifundia was usually merely a stage in the demographic cycle, representing a period of limited personal resources and dependence on wages. With savings, family inheritance, and some time, they would generally evolve into family farms. This was no longer to be the case for most minifundia as the land market had become almost inaccessible to individuals with limited means. Wage employment thus became important throughout the lives of minifundistas, at least until their children were grown and could help with family support.

TYPICAL MINIFUNDIA FARM HOUSEHOLDS: CASE STUDIES

Our two case study minifundia provide examples of a younger farmer, Sr. Galvéz, still tied closely to the "family farm" of his father; and an older farmer, Sr. Alemán, who was helping his own sons seek their way in a land-short world. Tables 6.1 through 6.3 summarize farm production and cost data from the case study field analyses.

Galvéz Family

Sr. Galvéz, 20, came from a family of ten children, all of whom lived nearby, whose father had five hectares of cacao. He was married to 20-year-old Adelina, daughter of a cacao producer and had two children by her. In 1973, his father bought him two hectares of grazing land that were being utilized for pasturing the father's cattle.

In 1975, Sr. Galvéz bought the two hectares of land on which the family resided. It was already planted to cacao, but in poor condition. He replanted 200 trees in 1978. In addition, he planted four "manos" of maize in Pechucalco ranchería, on land belonging to his father. Land, buildings, and equipment on the Galvéz farm were worth M$270,000-500,000 (over US$10,000) in 1979. But there was very little nonland capital or livestock. Gross farm labor earnings were $15,972,

Table 6.1 Inputs for Minifundia Cacao Farms, 1979-80

	Galvéz Farm	Alemán Farm	*Recommended*[a] w/o Rehab	*Recommended*[a] with Rehab
Productivity				
Number of trees	1,250 (625/h)	1,200 (1000/h)	600/h	600/h
Age 1	2 yr - 250	8-10 yr - 700		
Age 2	10 yr - 1000	20 yr - 500		
Age 3	--	--		
PDN/tree	.283 kg	.376 kg (.452 for trees in full production)	Average 1.3 kg[b]	
PDN/hectare	170.2 kg ($8,500)	301.4 kg ($15,070)	637 Average 5.38 kg[b]	835 kg/h
Man-Days/hectare	34-40	50	81.5	130
Pruning, clearing	13	2	21	24
Drains	1	29 (and shade)	3	6
Fertilizer	1	10-20	2.5	10
Insecticide	1	6-11	12	20
Fungicide	1.5	6-11	15	30
Harvest & Pod-breaking	16.5 (or 22, if divided by 1.5)	5.5	28	40
Inputs/hectare				
Fertilizer:				
Urea	--	--	--	2
18-9-18	--	--	--	--

15-15-23	--	--	--	--
Floren	2	2	--	1-2
Nutrisol	--	--	--	--
Gro-Green	--	1	--	--
Crescal	--	1	--	--
Insecticides:				
Trident	--	2	--	--
Metílico	2	--	--	2-3
Foley	--	--	--	--
Folidol	--	--	--	--
Folimac	--	--	--	--
Polidor	--	--	--	--
Fungicide:				
Lime & Sulfate	1	4	--	5-8
Cuprasol	--	--	--	--
Pruning	1	1	--	2-3

Note: kg = kilograms; h = hectare.

[a]This data is from estimates in the SARH, *Plan de Cacao*, 1979.

[b]This data is from the 1978 Cacao Census.

Source: Compiled by the author, Cacao Farm Budget Studies, 1979-81, Tabasco, Mexico.

Table 6.2 Costs of Cacao Production on Minifundia Cacao Farms, 1979-80

Input	Galvéz Farm	Alemán Farm		*Recommended*[a] *Plan*	
			Grijalva[b]	w/o Rehab[c]	with Rehab[d]
Labor	\$ 2,700	\$ --	\$ 9,400	\$ 8,100	\$13,900
Interest	--	200	--	--	--
Inputs	--	--	5,075	2,547	4,731
Miscellaneous	--	--	--	--	--
Fertilizer	--	--	3,200	1,500	2,366
Urea	--	--	--	--	--
18-9-18	--	--	--	--	--
Floren	720	--	--	--	--
Gro-Green	--	120	--	--	--
Nutrisol	--	--	--	--	--
Insecticide	--	--	675	1,047	2,365
Folimac	--	--	--	--	--
Tridente	--	200	--	--	--
Metilico	480	--	--	--	--
Methyl	--	--	--	--	--
BHC	--	240	--	--	--
Foley	--	80	--	--	--
Polidor	--	--	--	--	--
Folidol	--	--	--	--	--
Fungicide	--	--	1,200	--	--
Lime	5	--	--	--	--
Sulfate	40	200	--	--	--

Cash Costs	3,945	$ 1,040	$14,475	$10,647	$18,631
Depreciation	168	382	--	--	--
Full Costs	4,113	1,422	14,475	10,647	18,631
Equity Deducted[e]	28,000 (14,000)	14,000 (7,000)	14,000 (10,500)	14,000 (10,500)	14,000 (10,500)
Costs with Equity Deducted	32,113 (18,113)	15,422 (8,422)	28,475 (24,975)	24,647 (21,147)	32,631 (29,131)
Number of hectares	2	1.5	1	1	1
Total Cash Costs/hectare	5,700	21,440	--	10,647	18,631
Family man-days/hectare	--	--	--	--	--
Imputed cost family labor	4,822.5	14,293	--	--	--
Cash for inputs[f]	622.5	560	--	2,547	4,731
With depreciation[f]	2,056.5	948	--	10,647	18,631
With equity deducted[f]	16,056 (or 9,056)	10,281 (5,614)	--	24,647	32,631

[a]The original figures valued a man-day of labor at $75 (Grijalva data) and $83 (Cacao Plan data). Here we have adjusted this to $100, the minimal going rate found in Tabasco during my field study. At the more typical rate of $120, the estimates would have been substantially higher.

[b]SARH, Comisión de Grijalva, Dirección de Estudios y Proyectos, "Rendimientos y Costos de Producción de Cultivos Tropicales," August 1979, pp. 10-16.

[c]SARH, "Plan de INcremento de la Producción de Cacao en el Estado de Tabasco," January 1979, Villahermosa, Tabasco, Cuadro 20. This represents the supposed "typical" production system in Tabasco before rehabilitation programs.

[d]Same source as above. This represents the production costs after farmers have instituted rehabilitative measures on their plantations (see text) suggested by the Cacao Plan.

[e]A higher estimated value for land is shown in the first figure; a lower estimate in parentheses.

[f]These include imputed cost of family labor.

Table 6.3 All-Farm Production Costs on Minifundia Cacao Farms, 1979-80

Inputs	Galvéz Farm	Alemán Farm
Labor	$ 3,200	--
Cacao	2,700	--
Cattle	--	--
Maize	500	--
Cacao chemicals and miscellaneous	1,245	840
Vehicle[a]	--	--
Cattle Feed, etc.	1,560	4,320
Veterinary services and medicines	--	--
Equipment	--	500
Cattle purchases	--	--
Interest on loans	--	200
Rent land, equipment	--	--
Miscellaneous[b]	--	--
CASH COSTS	6,005	6,040
Depreciation	348	752
Cacao	--	--
Cattle	--	--
Maize, etc.	--	--
Presents of products	--	--
FULL COSTS	6,353	6,792
Equity capital costs	17,500-35,000	9,590-17,640
Cacao	14,000-28,000	--
Cattle	--	--
Maize, etc.	--	--
COSTS with deduction of 7% on equity capital	30,206-47,706	16,382-24,432

[a]This value includes tractors, trucks, etc.

[b]F. = fuel; T. = transportation.

Source: Compiled by the author, Cacao Farm Budget Studies, 1979-81, Tabasco, Mexico.

78 percent of which was from cacao. Total production for 1979-80 totalled 281 kilos of dried beans.

In no month did the Galvéz farm require a full man-month equivalent of labor. Yet Sr. Galvéz was fully occupied, as most of his time was spent working for pay for his own father. During peak periods in August, February, and March he worked very long days in both places, requiring help on the farm from young José Dolores, who was paid $100 per day. Adelina worked at home nearly all the time, rarely leaving even to shop. September through April she helped in the cacao harvest several afternoons and in the Saturday pod-breaking. Of total man-days worked on the farm (80), Sr. Galvéz provided 60 percent and José nearly all the rest.

Table 6.4 Minifundia Cacao Farm Household Gross Income, 1979-80

Source	Galvéz Family	Alemán Family
Cacao	$12,422	$ 25,892
Poultry	1,800	--
Coconut	--	696
TOTAL Farm Income	14,222	26,588
Farm produce consumed	2,375[a]	6,400[b]
Change in livestock inventory	-625	-500
Miscellaneous	--	--
TOTAL Farm Labor Earnings	15,972	32,488
Off-farm agricultural work	26,400[c]	13,920
Off-farm nonagricultural work	--	63,740
Loans	--	3,000
TOTAL Income	$42,372	$110,148

[a]Maize-1440; pig-500; poultry-1,875.

[b]Maize-1,800; hog-500; fruit-2,000; eggs-900; turkey-1,000.

[c]220 days @ $120.

Tables 6.4 and 6.5A show minifundia income data. For the Galvéz family, net cash income per family man-day of work in cacao was $204; per general man-day it was $131. If a deduction for equity was made, however, these values plummetted

Table 6.5A Family and Farm Income of Minifundia Cacao Farm Households, 1979-80: Galvéz Family

Income Measure[a]	Cash Income[b]	Results: Labor Earnings[c] without Equity Return Deduction	Labor Earnings[d] with Equity Return Deduction
a. Net farm __	$ 8,217.00	$ 9,619.00	$ -31,734.00 (-14,234)
b. Net off-farm agricultural __	26,400.00	26,400.00	26,400.00
c. Net nonagricultural __	--	--	--
d. Net total __[e]	34,617.00	36,019.00	-5,334.00 (12,166)
e. Net farm __ per hectare	2,054.00	2,404.75	-7,933.50 (-3558.50)
f. Net farm __ per man-day on farm[f]	103.35	120.99	-399.17 (-179.04)
g. Net farm __ per family man-day on farm[g]	172.99	202.50	-668.08 (-299.66)
h. Net nonagricultural __ per family man-day worked outside agriculture[g]	--	--	--
i. Net agricultural off-farm __ per family man-day worked[g]	120.00	120.00	120.00
j. Net total __ per family man-day worked[g]	129.41	134.65	-118.63 (-53.21)
k. Net cacao __ per man-day in cacao[f]	131.42	128.82	-305.29 (-88.23)
l. Net cattle __ per man-day in cattle[g]	--	--	--
m. Net pig __ per man-day in pig work[f/i]	--	--	--
n. Net __ __ per man-day in __[f]	--	--	--

o. Net cacao __ per family man-day in cacao[g]	204.26	200.21	-474.48 (-137.13)
p. Net cattle __ per family man-day in cattle[g]	--	--	--
q. Net pig __ per family man-day in pigs[g]	--	--	--
r. Net __ __ per family man-day in __[g]	--	--	--
s. Percent labor in cacao expenses[h/j]	68.4	65.6	8.4
t. Percent labor in whole farm expenses[h/j]	53.3	50.4	6.7

Note: The value in parentheses in the last column reflects lower estimated value of land resources. Irregularities in the Galvéz budget include: (a) uncertain fungicide use; (b) unknown the exact number of days in which he worked for his father; (c) uncertainty about the level of maize purchases in 1979-80.

[a]The blank spaces below refer to the specific income measure used--cash and labor earnings with and without equity deduction; the figures are found in the appropriate column to the right.

[b]Cash income refers to costs and returns to the family only in cash.

[c]"Labor Earnings" (without equity deduction) is a measure of income that includes several noncash components such as: inventory changes, family use of farm products, etc.

[d]"Labor Earnings" with equity deduction is similar to the simple labor earnings measure, but it also adds the cost of tying up the family's capital in the farm enterprise. It enters, as a cost, the interest that could be paid to the family by a bank were all assets to be sold for cash and deposited,

Table 6.5A Continued

without the risks entailed in a farm enterprise. The very low rate of seven percent is used for our calculation.

[e]Total Income of whatever measure is the sum of farm income, off-farm agricultural income, and off-farm nonagricultural income.

[f]Income per man-day is a rough measure of average labor productivity including both hired and family labor use. To compute this, the wage bill is subtracted from total costs.

[g]This is the average value of a day of labor performed by family members. For this purpose, the wage bill *is* included in costs.

[h]The value in parentheses represents the higher of two cost estimates for cacao production.

[i]This pig budget is probably inaccurate. If we assume net returns at $1 million (based on unreported costs) and a doubling of the very low reported labor input to 800 man-days per year, then returns per man-day are reduced to $1,250.

[j]This does not include the imputed value of family labor.

[k]The value in parentheses represents costs (or returns) in the form of a changed animal inventory.

Source: Compiled by the author, Cacao Farm Budget Analyses, 1979-80, Tabasco, Mexico. Note that the costs, returns, and labor use summarized here can be found in preceding chapter tables.

into the red. Cash income for the farm as a whole was $173 per family man-day and $103 per general man-day. Nevertheless, total cash income per family man-day worked, including non-farm sources, was only $129--approximately the going local wage--because of Sr. Galvéz' work with his father. Immediate returns from this latter work were not competitive with returns for work at his own farm.

Sr. Galvéz experienced cash deficits in April, June, July, September, and December, but they were not major and were interspersed between surplus months. This was made possible by his access to unlimited employment opportunities with his father to make up any deficit. A subsistence level income estimated at $25,000-30,000 for his family--or $2,100-2,900 per month--could be earned in 18-25 days at $120 per day of full or near-full off-farm employment. Farm income was largely supplementary.

Alemán Family

Sr. Alemán, 54, had a one-and-a-half hectare cacao farm in Huimango II. He also rented three "manos" of milpa nearby "al tercio"--by working the land, he received one-third of the harvest. He lived on the farm with his wife Josefa, 52, and seven of their children, aged 6-24.

Another three children, aged 18-26, were married and lived nearby. The children over 20 had studied only through third grade; the five younger ones were in primary and secondary schools. Three sons, two married and one unmarried, were masons. Another unmarried son was a carpenter and another had his own cacao plantation. The married daughter had formerly been a nurse, but in 1980 was a full-time housewife.

Sr. Alemán's father was a jornalero who, through savings, managed to acquire four hectares of land when his son was 12. When Sr. Alemán married at age 22, this land was divided between him and a brother, and the father bought three hectares elsewhere, where he resided. The Alemáns sold one "rial" (10 "manos") fifteen years ago, but otherwise there had been no land use changes.

His capital was worth between M$137,000 and $252,000 in 1979, depending on land valuation. The value of his nonland capital was six times greater than Sr. Galvéz'. Gross total farm labor earnings were $32,488, of which 80 percent was from cacao sales and almost 20 percent from family-consumed farm production. Total cacao production for 1979-80 was 469 kilos of dried beans.

Sr. Alemán worked primarily in cacao; he no longer worked as a jornalero, which in earlier years had kept him busy. The others did cacao work on weekends. Yet of all the chil-

Table 6.5B Family and Farm Income of Minifundia Cacao Farm Households, 1979-80: Alemán Family

Income Measure[a]	Cash Income[b]	Results: Labor Earnings[c] without Equity Return Deduction	Labor earnings[d] with Equity Return Deduction
a. Net farm __	$20,548.00	$ 26,248.00	$ -15,218.00 (2,282)
b. Net off-farm agricultural __	13,920.00	13,920.00	13,920.00
c. Net nonagricultural __	63,740.00	63,740.00	63,740.00
d. Net total __[e]	98,208.00	103,908.00	62,442.00 (79,942)
e. Net farm __ per hectare	10,274.00	13,124.00	-7,609.00 (1,141)
f. Net farm __ per man-day on farm[f]	156 [or 120]	200 [152]	116 [88] (17[13])
g. Net farm __ per family man-day on farm[g]	156 [or 120]	200 [152]	116 [88] (17[13])
h. Net non-agricultural __ per family man-day worked outside agriculture[g]	166.80	166.80	166.80
i. Net agricultural off-farm __ per family man-day worked[g]	102.30	102.30	102.30
j. Net total __ per family man-day worked[g]	150.3 [or 141.2]	159.12 [149.51]	95 [89] (122[i][114])
k. Net cacao __ per man-day in cacao[f]	241 [173]	196 [150]	79 [61] (133[i][102])
l. Net cattle __ per man-day in cattle[g]	--	--	--
m. Net pig __ per man-day in pig work[f/i]	--	--	--

n. Net __ ___ per man-day in __[f]	--	--	--
o. Net cacao __ per family man-day in cacao[g]	241 [173]	196 [150]	79 [61] (133[i] [102])
p. Net cattle __ per family man-day in cattle[g]	--	--	--
q. Net pig __ per family man-day in pigs[g]	--	--	--
r. Net __ ___ per family man-day in __[g]	--	--	--
s. Percent labor in cacao expenses[h]	0	0	0
t. Percent labor in whole farm expenses[h]	0	0	0

[a]The blank spaces below refer to the specific income measure used--cash and labor earnings with and without equity deduction; the figures are found in the appropriate column to the right.

[b]Cash income refers to costs and returns to the family only in cash.

[c]"Labor Earnings" (without equity deduction) is a measure of income that includes several noncash components such as: inventory changes, family use of farm products, etc.

[d]"Labor Earnings" with equity deduction is similar to the simple labor earnings measure, but it also adds the cost of typing up the family's capital in the farm enterprise. It enters, as a cost, the interest that could be paid to the family by a bank were all assets to be sold for cash and deposited, without the risks entailed in a farm enterprise. The very low rate of seven percent is used for our calculations.

Table 6.5B Continued

[e]Total Income of whatever measure is the sum of farm income, off-farm agricultural income, and off-farm nonagricultural income.

[f]Income per man-day is a rough measure of average labor productivity including both hired and family labor use. To compute this, the wage bill is subtracted from total costs.

[g]This is the average value of a day of labor performed by family members. For this purpose, the wage bill *is* included in costs.

[h]The value in parentheses represents the higher of two cost estimates for cacao production.

[i]This pig budget is probably inaccurate. If we assume net returns at $1 million (based on unreported costs) and a doubling of the very low reported labor input to 800 man-days per year, then returns per man-day are reduced to $1,250.

[j]This does not include the imputed value of family labor.

[k]The value in parentheses represents costs (or returns) in the form of a changed animal inventory.

Source: Compiled by the author, Cacao Farm Budget Analyses, 1979-80, Tabasco, Mexico. Note that the costs, returns, and labor use summarized here can be found in preceding chapter tables.

dren, only Santo (the carpenter) and Salomón (part-time cacao worker, part-time mason) seemed to work full-time. Only a quarter of all family work for cash income was directed to the farm.

Josefa used to be active in cacao production. But since her older daughter married in 1978, she had been overburdened with household work: she cooked, washed, and cleaned for four working men and three students, with only occasional help from her sons.

Of total man-days worked on the farm (171), 54 percent was provided by Sr. Alemán and the rest by his sons. The net cash farm income per man-day worked was $120-156; $173-241 per man-day in cacao; $141-150 per family man-day for all activities, including off-farm agriculture and nonagricultural work ($102 and $166, respectively). (See Table 6.5B.) It was not clear why more time was not spent on the farm relative to off-farm agriculture. It may have been because cacao income was for the most part shared among the family, while the sons' off-farm work was claimed primarily by themselves.

Perhaps cash flow constraints were to blame. Total income varied from nearly $15,000 in December to $5,000 in May, although average monthly family and farm expenses were $8-9,000 (including sons' personal expenses). February through July there was a cash deficit, only partly relieved by a $3,000 personal loan received in February.

Summary of Case Studies

Cacao yields for both Sr. Galvéz and Sr. Alemán were well under state averages, even though they used recommended levels of foliar fertilizer and insecticide. Sr. Galvéz used an unusually low level of fungicide. Input costs per hectare were only a quarter to a fifth those of Plan "traditional" levels; only 1/7-1/9 of the "recommended" level.

The most notable departure of cacao production systems from both "traditional" and "recommended" systems was the far lower labor input, most surprisingly for those items requiring least associated cash input: fungicide application, pruning, desmamone, and drain and shade control. This was so despite the fact that minifundista farmers rarely paid family members for their share of cacao work, beyond room and board. Even when the head of household worked full-time off-farm and installed his son as main producer, the son was usually paid a share of the proceeds, rather than a daily wage.

Whole farm costs were considerably above cacao enterprise costs, due to the large amount of purchased feed for domestic livestock. As this was mainly for their own use, it should perhaps be accounted for under consumption rather than production. Depreciation costs were negligible because of the lack of fixed capital; but costs with equity return deducted were 1.5-5.0 times greater than full costs.

Due to Sr. Galvéz' hiring labor, his cash costs were nearly four times those of Sr. Alemán, for only a third more cacao land. However, if we add imputed family labor costs at $120 per man-day, then Sr. Aléman "spent" twice as much as Sr. Galvéz, and nearly three times more per hectare of cacao; his yields were then 77 percent higher per hectare.

Note that these two families were at points in their life cycle when the home farm held especially low priority among the family income-producing options. The yield evidence from the CCT showed higher yields for minifundia than for larger holdings, probably because most minifundia were in a demographic stage somewhere between that of Sr. Galvéz and Sr. Alemán's families, when significantly more family labor was applied to cacao.

MINIFUNDISTA FAMILY ECONOMIC STRATEGIES

The basic objectives of the minifundista family seem to be the following: (1) maintenance of long-term income at or above subsistence, and reduction of risk to that income; (2) use of land as a family economic and social "base;" and (3) maximization of returns to labor from all sources.

Subsistence Concerns

The case study family consumption budgets suggested how close to subsistence these families live. The hospital visit by Adelina could only be paid for by her father-in-law's generosity, not the family's own resources. Fifty to sixty percent of the budget was spent for food. Thus, we can see the importance of on-farm production. Were this included in the consumption budget, the proportion spent on food would be still higher.

A continuing balance between on-farm subsistence production, farm production for cash, and off-farm cash employment was made by these families to keep real income--and particularly food consumption--above the subsistence line. Not unexpectedly, the main evidence of an improved standard of living among minifundia households was a more diverse diet and an increase in meat consumption to once or twice a week in recent years.

In the SST sample, 84 percent of minifundia had cacao as the primary product; an additional 11 percent had maize or coconut as the principal crop. (There were more maize and coconut farmers among minifundistas than any other group). Because of space limitations and the recent option of a high value cash crop, over a third of the farms had no product other than cacao. A fifth had maize as a secondary crop. Many more planted small amounts of maize as a tertiary product, but this was not picked up in the SST.

Subsistence concerns were particularly important, given the demographic composition of most minifundia households. Despite fewer resources than family farm households, the dependency ratio was higher, imposing greater burdens on workers. Minifundia also had female heads of household somewhat more frequently than did other family types. But overall, female economic participation rates were low, because so many were mothers with many young children. Half of those women who did work were heads of household.

Land as a Family Economic and Social Base

This generation of minifundistas did not view the land as marketable. It was not capital and not a redeemable asset, except under the most extreme of economic emergencies, when land had to be sold for lack of any other recourse. Nor could they hope to increase their holdings, with cacao land costing $200,000 per hectare. "Returns to the land" or "capital" was therefore not included in their economic calculations. Estimates of labor earnings with equity deducted were simply not relevant to their calculus, and they were undisturbed by "negative" incomes as measured by this criteria.

Apart from its role as security, the land provides the potential to revert to home production of a considerable part of the family food needs in case of economic upheaval or crash in the cacao market. Most individuals were aware that the land could not provide the principal economic support for the family. It was certainly not a form of long-term livelihood for younger members of the family. Land could not be passed on to a child for use before the parents' death unless the parent had alternative retirement support, which was very unusual. Thus, the heir's future was precarious, not to mention that of siblings. Where parents felt obliged to subdivide the land further, the resulting fractions were truly only marginal to the family economy.

But the land had another role as well. There was extremely high job mobility among the younger generation of cacao producer families, though mobility was of a different type than their parents'. Few individuals had a history of permanent migration, according to the SST; those who did were largely over forty years old. (That age group had experienced more migration than those of any other farm type.)

The younger generation nearly always returned to settle near home--for cultural reasons: the lower cost of raising very young families under an in-law's roof; the lower cost of building a house on family property; and the availability of family help in erecting a home. As off-farm income increased, and the variety of income sources increased, many of these farms became more important as a home base for circular migration than as a productive unit (Bromley and Grey 1979, pp. 17-19). Even the tiniest cacao minifundia were important for population settlement.

The Role of Off-Farm Employment

In many agricultural studies, particularly those of the "proletarista" school and many neo-Marxists, as well as more orthodox economists, off-farm employment by farmers has often been associated with a trend to proletarianization, and a decline in farm production and productivity (e.g., Griffin, in Edwards 1968, p. 198; de Janvry and Deere 1978, p. 19; Barkin and Suárez 1982).

This interpretation of the role of off-farm labor is misleading for the case of Tabascan minifundia producers. First it assumes that the replacement of a subsistence crop by a cash crop is a sign of the breakdown of the subsistence orientation of the family. It next assumes that off-farm employment is a sign of deteriorating economic conditions for the family and that off-farm employment is necessary because of a deterioration in the ability of on-farm production to provide subsistence requirements.

Unlike all hired-labor-based producers and some family-labor-based farmers, minifundista cacao producing families have maintained and for the foreseeable future will maintain a subsistence-oriented economic strategy; they must, because of the scarcity of their means. But the very size limitations that define them make an autarkic "subsistence" farm impossible. Indeed, the true, diversified, self-sufficient "subsistence" farm of the literature was closer to the family-labor-based farm before the 1970s, described in Chapter 5.

The minifundia family of today must guarantee both a minimum food supply and a minimum cash supply for nonfood and bought-food needs. Trade-offs between farm maize and cacao production on the one hand, and between cacao production and off-farm employment on the other, have been made in this context for decades. The outcome depended historically in part on relative crop prices, in part on the availability of off-farm income supplements, and in part on the on-going historical increase in area of established cacao plantations. Family labor power could be re-allocated easily, resulting in major shifts in farm cost and income structures.

Diversity of income sources was also a means of spreading risks, although the short-term noncontractual nature of most off-farm jobs meant this source was not always dependable, and valuable time was lost in searching for jobs. Sons were far more willing to do this than were their fathers, because few of them foresaw an economic future on their father's farm. While such farmers were unable to prevent the proletarianization of some offspring, the units themselves were viable for the long term.

Effect of Off-Farm Employment on Farm Production. Off-farm employment did not usually imply an abandoned farm. As Martínez and Rendón asserted (1978, p. 669):

> *Forgetting that real income is not always equal to the cost of farm reproduction and that paid work may fulfill different functions according to how different types of campesinos carry them out, leads to the erroneous conclusion that the greater the quantitative importance of wage income, the greater the tendency to proletarianization [my translation].*

Wage labor may even be complementary with on-farm production. Sr. Alemán, for example, had a long-term relationship with a neighbor whereby he sprayed her cacao trees and received, in addition to a wage, free use of her sprayer for his own trees. When she died and he couldn't find a similar arrangement, he went to work in PEMEX construction for a year to make enough money for his own sprayer, at which point he returned to full-time farm work.

This utilization of nonfarm employment for specific cash goals was very widespread. Repayment of debt was a common reason, as was purchase of livestock. Farmers' sons worked off-farm to earn cash to set up a carpentry shop or pay for an apprenticeship; to assert independence or learn to manage their own budgets; or because their parents were unwilling or unable to continue to support them.

Off-farm employment did not necessarily result in less attention to the cacao plantation. With few trees, families at certain demographic stages could readily replace much of the lost labor. This was so particularly when children were in their early to mid-teens: the mother was less occupied with infant care and sons were not yet anxious to earn money off-farm, or indeed may still have been in school or apprenticeship.

To the extent that off-farm employment is used to meet critical cash constraints, higher wages and increased off-farm employment opportunities may even increase the number of days available to the farmer for working in cacao, or his level of cash investment in the farm.

Implications for Cacao Production. Both cash and labor were important constraints to increased cacao production on minifundia farms, and these factors were intimately intertwined.

Let us assume that the "recommended" techniques of the Cacao Plan were applied to the 1.5 hectares that both Sr. Galvéz and Sr. Alemán had in production. If they had been able to achieve yields of 825 kilos per hectare by using $5,000 per hectare of purchased inputs and all family labor, total family cacao income would be $54,375--four times the 1979-80 income of Sr. Galvéz, twice that of Sr. Alemán.

Yet this would have required a doubling or tripling of family labor use on the farm and an 8-10 fold increase in cash costs. This income would be delayed until harvest and would carry enormously greater risk. And if the extra 59-90 days of work required for intensification had instead been spent in off-farm employment, the family would have earned far more. Instituting the Cacao Plan recommendations on the "typical Tabascan farm" required an extra 57.5 days of labor, resulting in additional cacao income of $1,015. If the same number of days were worked off-farm, even at the low reference wage of $83, income would increase by $4,772. At 1980 wage levels of $120, the difference would be even greater; and minifundia farmers needed to work an additional 58, not 90, days on their farm in order to reach recommended levels.

Since family savings were usually earmarked for long-term investment or large consumption items, cash for purchase of new inputs which markedly increase productivity had to be acquired elsewhere. The risks were substantial. Sr. Alemán, for example, though he had the opportunity to borrow $18,000 from the bank for 1.5 years to cover cacao costs (his credit was very good), was too afraid of the consequences of a poor harvest and future debt. On the other hand, it did not make sense to work an extra forty days off-farm specifically for purchase of additional cacao chemicals. For the $5,000 in chemical costs, there had to be added the opportunity cost for the extra sixty days of work required to apply the inputs. Yields would have had to increase 340 kilos per hectare just for the farmer to break even.

Sr. Alemán instead borrowed $3,000 as a personal loan during his low income months, mainly in order to purchase chemicals December through March with his income from the harvest. It was preferable to raise investment funds through a small loan; interest payments were far less than the opportunity cost of the extra labor.

Statistical Employment Trends

Chart 6.1 provided data on the employment characteristics of the minifundia, which largely corresponded to the strategies sketched above. It showed a high percentage of farmers with their principal occupation off-farm. A quarter of these farmers had a secondary occupation, most commonly as jornaleros.

Minifundia owners were more likely than other farmers to work in skilled trades and construction. Minifundia families had the highest overall employment in unskilled work and very high participation in nonagricultural work (20 percent of males). Because of the importance of stable off-farm employment, they had the lowest proportion of agricultural workers.

Only ten percent of those who did work as jornaleros were heads of household. Of these workers, 84 percent were under 30 years old; most were under 20. On the other hand, a high proportion of individuals over 50 years old were jornaleros, relative to other cacao groups.

Nearly a quarter of the EAP ever in cacao had left cacao work. Nearly a third of these were no longer working; a third were in noncacao agricultural work; and nearly a third in skilled trades.

Between 1965 and 1978, there was little change in the number of minifundia farm owners working outside agriculture as their principal occupation, but between 1978 and 1979 there was a big increase. But there was no change between 1965 and 1979 in the proportion of minifundistas whose principal occupation was agricultural worker.

MINIFUNDISTA RESPONSE TO ECONOMIC CHANGE

The response of minifundia farm families to changes in input subsidies, cacao prices, and wages differed from the other two types of producers, primarily due to the importance of off-farm wage employment in the family strategy.

As cacao prices increased, farmers planted land formerly in maize to cacao, and increased input applications. Land availability was the main factor restricting total production as the price rose. Minifundia were capable of greater labor intensity and hence higher productivity because of the higher average worker:land ratio, but this was so only at stable wages lower than the value of a family man-day in cacao. Otherwise, family farms would generally outproduce them.

If the real price of cacao were to go down, there would be a doubly negative effect on minifundia family workers. To maintain incomes, they would have to increase work off-farm; this very addition to the labor force, coupled with reduced labor demand from larger-scale cacao producers, would be likely to reduce the average wage and further increase the number of days that had to be worked to maintain the standard of living.

But the cacao plantations would not be abandoned until the price went considerably lower than the level at which other producers would stop, particularly if the wage level declined, leaving few alternative sources of cash.

The wage response of minifundistas was complicated, as implied above. But there were several clear differences between minifundistas and other types of producers. First, unlike on the family farm, the labor use function did not include the number of family members available to work (since there were by definition enough), and did not include the hiring of farm workers.

The family standard of living was always improved by an increase in the wage rate. All increases in cacao production were associated with more labor than cash (chemical) inputs, although the rate would depend on the ratio of workers to dependents, as it did for family farms.

If the wage increased, there might be a decrease in the number of days worked off-farm to meet specific cash needs, and a corresponding increase in labor on the farm at least for the principal farmer. Younger sons might still prefer to work with "faceless" corporations or as city employees, where they have more independence and status, rather than work under an overseer on ranches and large cacao farms. If the availability of off-farm, nonagricultural employment were to decrease (or the wage decrease), workers would return to the hired-labor-based employers, if they still needed cash, and intensify production on their own cacao plots. Under other circumstances, decreasing wage levels may result in reduced commercial farm production among these families, as the family was forced for reasons of security to revert to greater subsistence production on the farm.

Should both off-farm wages and the price of cacao decline, the minifundista may revert to old subsistence patterns, with greater emphasis on maize, greater sibling rivalry over who gets access to the plot, and the consequent likelihood of increased permanent as opposed to temporary migration.

AGRICULTURAL PROGRAMS FOR MINIFUNDIA CACAO FARMERS

Increased cacao prices led minifundistas to adopt foliar fertilizers and increase insect and fungal control to an extent unappreciated by most technical cacao personnel. However, levels used were pitifully low due to the cash constraint on production. For financial reasons, the Cacao Plan program was simply out of reach for most of these producers--both the costs of inputs and the opportunity cost of family labor.

The program of producer cash loans needed to be adjusted substantially for the minifundistas. They could not responsibly take on the large sums loaned for cacao production by the banks. But loans for much smaller amounts, perhaps distributed by the cacao associations, could be very useful, and would encourage greater production and input use by removing cash constraints during periods of input applications or plantation maintenance. These sums should not be restricted to use directly in cacao production, but be applicable to personal needs as well.

The practice of providing some input subsidies was effective for this group of farmers. More cacao associations could be encouraged to provide inputs on loan, their costs to be deducted from the pay at harvest time. But loans made entirely in the form of inputs would often lead to the family's selling off inputs to meet consumption needs at critical times.

Technical assistance programs need to adjust the timing of visits and presentations to permit cacao producers who are also full-time off-farm workers to participate. Similarly, it is important for technical personnel to recognize the variation of the production system over the life-cycle of the family, which is more pronounced among minifundia than for other groups.

In the long run, the coupling of high cacao prices and ample nonagricultural employment, particularly in skilled trades, could slow down the subdivision of agricultural land. The State Plan and the Cacao Associations should both encourage such training of farmers' children, and adjust cacao promotion programs so that minifundia can participate fully. Policies to reduce land subdivision among private property owners, and reinforce ejido land transfer regulations, are needed to stem the general process of minifundization in the state.

THE ROLE OF MINIFUNDIA IN A PETROLEUM ECONOMY

Minifundia in Mexico, and many other parts of the world, tend to be basically oriented to subsistence production; the small home plot provides a source of maize and beans for the family while wage labor is utilized to provide other consumption items. The plot rarely covers total consumption requirements, but rather acts as a sort of insurance policy for the family. Under conditions of severe economic stress, such families tend to increase the search for wage labor and other cash strategies, intensify subsistence production, or both.

In Tabasco, cacao, which had long been an integral part of the Tabascan subsistence farmers' activities, became transformed into a cash crop more valuable than ever before. Minifundia response to this situation, in the midst of an unprecedented wage labor market, provides a very provocative case study of the production potential of the minifundia. It brings to the fore questions of precisely where the weaknesses of the minifundia mode of production lie--whether within the family structure, the farm structure, the relationship with the wage economy, or relative prices. And while the Tabascan experience suggests that there are severe constraints on programs of technification which require large cash inputs, it also suggests the potential for stable and rising agricultural production by minifundia even under the stressful conditions of a petroleum boom.

The most important effects of the petroleum boom on minifundista families in Tabasco were sharp increases in cash needs due to inflation, and in the number and quality of jobs available for minifundistas outside agriculture. In the short run, the availability of petroleum-related employment had not reduced cacao production by this group of farmers, because of the seasonal nature of the jobs, and because most families could easily find substitute or assistant cacao labor. Cash transfers from petroleum and construction work for investment in cacao sprayers or inputs were widespread in what was basically a complementary relationship. The minifundia also played a critical role in retaining population in the countryside and slowing rural-urban migration.

Continuing subdivision, however, threatened to turn those plantations into mere garden plots. While such a system could come to resemble the lucrative, intensive Japanese garden plots tended by families of city workers, it could also lead to the stagnation of commercial cacao production over a large area of the state. Under conditions of either low prices and wages or high prices and wages, there will be a tendency for minifundistas to maintain the farm as a central core of family economic strategy. But a prolonged period of low agricultural prices and high off-farm wages could indeed accelerate the process of proletarianization of the countryside in petroleum economies.

7

Collective Farm Production Strategy

Collective ejidal production in Mexico dates back to some of the earliest ejidos organized in the country in the 1930s. Collective organization was seen by many as the only practical way to restructure the large private holdings that had been expropriated complete with central machinery plants or processing units. Others in favor of collectivization stressed social continuity with indigenous collective production forms; still others supported collectives on the basis of socialist ideology.

A considerable literature on collective production in Mexico has built up over the past fifty years,* and the fortunes and numbers of collective ejidos themselves have fluctuated considerably from one presidential sexenio to another. The periods of greatest support for them occurred during the Cárdenas years and more recently during the presidential term of Luís Echeverria Álvarez (1970-76). Collectivization received much less attention in the López Portillo and de la Madrid administrations.

In Tabasco, collective cacao production was responsible for a small, but rapidly growing part of total state production, with about eight to ten percent of the total area planted to cacao. It was an important source of ejidal employment and income.

Many issues related to collective farming have bearing on general agricultural development. Policies to promote farmer cooperation, credit societies, rainfed and irrigated agricultural district organizations must deal with similar

---

*Particularly good references on collectivization in Mexico are Eckstein (1966) and Reyes, Osorio, Stavenhagen et al. (1974).

problems, particularly the proper degree and type of government intervention, maintenance of democratic decision making, efficiency of bureaucratic decision making, and farmer organization, morale, and incentives. The case study of Ejido C-29 of the Chontalpa Plan, and the evolution of cacao production there, provide a practical look at the working out of these problems. The advantages and disadvantages of being part of a major governmental development scheme in a petroleum region also highlight some of the political choices that can or should be made in resolving agriculture-petroleum conflicts.

THE CHONTALPA PLAN

Nearly all of the collective cacao production in Tabasco was located in the "Chontalpa Plan" ejidos, an area drained in the 1960s as part of the state's first agricultural project under the auspices of the Grijalva River Basin Commission. This commission was organized in June 1951; in 1960 work was begun on the Netzahualcoyotl Dam, designed to protect Villahermosa and the central Chontalpa area from flooding.

The original "Plan Limón" was organized in the early 1960s to increase agricultural production in a major subsistence area and to increase real incomes and living conditions for smallholders through drainage and irrigation on about 40,000 hectares. In 1965, an enlarged project called the "Chontalpa Plan" was funded by the Inter-American Development Bank for 6,000 families. The following year, the Commission began to expropriate land (eventually from nearly 7,000 individuals) for the project. This land was organized into new ejidos, based on individual plot control, and credit through "Solidarity Groups."

Drain construction was finished by 1970. A year earlier, production was undertaken by "Local Societies for Collective Ejidal Credit," and the 15 hectares of land earmarked for each ejidatario was henceforth divided into two hectares for a family private plot and 13 hectares for the collectively run plots.*

In mid-1972, the Fideicomiso Plan Chontalpa was set up to reorganize the ejidos financially, with the Grijalva Commission still charged with technical and social assistance. By 1975, over half the planted land was in pasture and a large part in cane. By this time, all 22 "poblados" had been

---

*Presidential Resolutions legislating collectivization were passed in 1971 for Cárdenas municipio ejidos and in 1973 for Huimanguillo.

set up, complete with paved roads, brick housing, bus service, medical care, and schools for ejidatario's children. Numerous professionals such as social workers and crop technicians were active in attempting to smoothe the transition from disperse subsistence production to highly controlled, collective "modern" agriculture and "urban" living patterns. Between 1959 and 1975, infrastructure investment in the Plan as a whole was M$1,150 million, or an average of $52.3 million per ejido (approximately US$4.2 million).

In 1977, another reorganization gathered all responsibilities for the collectives under a single "Promotora del Complejo Agroindustrial de la Chontalpa," which answered directly to the President. Its mandate was to increase production and industrialization and to work with the "Union of Ejidos" (the ejidatarios' organization), gradually including them in decision making until full self-management was achieved.

Land Use Changes in the Chontalpa Plan

Agricultural production on the Chontalpa ejidos was one of the few economic variables in the state over which the government had some direct control. Success in this project was desirable for several reasons: its potential role in resolving local agricultural shortages; its importance as a symbol of government competence; its role as an example for tropical agricultural development in Mexico; and, for its proponents, support for continued collectivization of agriculture.

Throughout the 1960s the project emphasized intensive agricultural production. When President Echeverria came to power, several factors led to a change in orientation: (1) national presidential priority for increasing ejidal yields; (2) a crisis in the national sugar sector requiring increased area in cane and improved productivity; (3) a national beef shortage arising from large-scale beef exports to the U.S.

Production was radically reallocated to pasturing cattle and planting cane, with the erection nearby of a cane mill. Herd size in the Chontalpa rose from 1,042 in 1969 to 41,930 in 1975. These decisions were reinforced by poor initial production results in agriculture and a belated recognition that Chontalpa soils, with the exception of some alluvion, suffered serious permeability problems. Given very high rainfall, tractor use was extremely problematic and was still a very high-cost input. Heavy losses in these years, due to extremely high rainfall and consequent flooding, finally led to the initiation of parcel-level drainage construction, which had long been needed for more precise water control.

In 1980 and 1981 there was again a push to produce subsistence crops, particularly maize, as part of the Sistema Alimentario Mexicano.

Chart 7.1 illustrates land use changes in the Chontalpa since 1965. Thousands of hectares of cacao were cut down in the early years when emphasis was on annual crop production, cacao prices were low, and division of land into large blocks of monoculture was technically appealing. Area in cacao dropped from 6,150 hectares in 1965 to a low of 2,535 hectares in 1971.

As cacao prices rose, communal and private plantings began anew, reaching 3,835 hectares in 1975. By 1979, there were 3,337 hectares on communal land alone; individual plot cacao plantings continued to increase. An additional 500 hectares were planted on collective plots 1979-80; about the same on private plots. Fifteen Chontalpa ejidos produced cacao in 1980; on ten of them over 100 hectares were in production or development. By the end of the sexenio, in 1982, the ejidos expected to have planted an additional 3,500 hectares.

History and Description of Ejido C-29

C-29, or as it was formally but rarely called, Ejido General Vicente Guerrero, covered approximately 5,000 hectares in the southeastern part of the Plan. Maps 7.1 and 7.2 show land use changes on the ejido between 1965 and 1980. The ejido was entirely agricultural, with no agro-industry but the washing and drying of cacao.

The zone was originally one of small ranches and minifundia, with at least 1,000 hectares planted to cacao on over half the individual holdings. In 1979 the ejido had 600 hectares of cacao plantations (two thirds on communal land), 80 of sorghum, 1,110 of sugar cane, and 850 of pasture. Its cattle herd numbered 1,660; 526 head were cows producing milk for Nestle's. The collective ejido had two entirely separate production systems for the communally-worked fields, and for the two hectares per family private plots.

By far the largest capital component of the ejido was its land, which did not legally belong to the ejidatarios. At 1980s' inflated prices, the land was worth $420 million pesos. Equipment was valued at under half a million pesos. The 160 ejidatarios with rights generally owned concrete block houses, but the "libres" and poorer ejidatarios without rights still had homes of guano.

In 1972, the ejidal population was estimated to be 700 individuals in 121 families (Barkin 1976, p. 145). The total population in 1976-77 was 1,220, of whom 650 were men and

Chart 7.1 Land Use Changes in the Plan Chontalpa, 1965–79

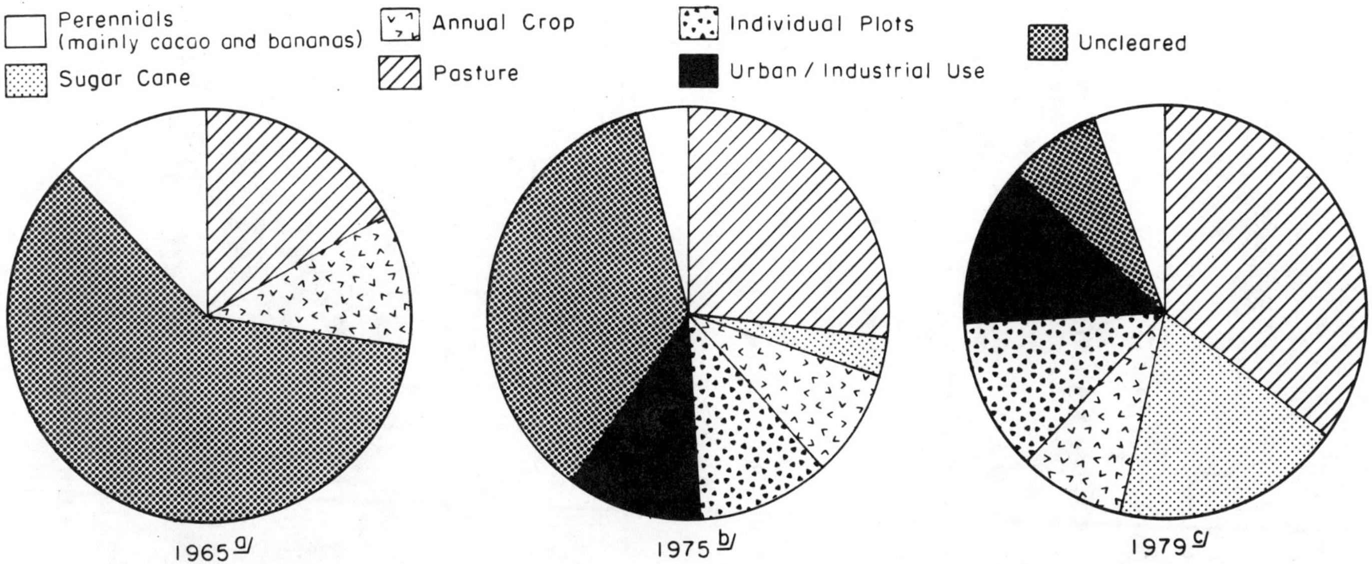

[a]Rosario Casco Montoya, *Manejo del Agua en un Ecosistema Tropical: El Caso de La Chontalpa*, Centro de Eco-Desarrollo: Mexico, 1979.

[b]A. René Barbosa-Ramirez, *La Ganadería Privada y Ejidal: Un Estudio en Tabasco*, Centro de Investigaciones Agrarias: Mexico, 1974.

[c]"Promotora del Complejo Agro-Industrial de la Chontalpa: Programa Para 1979" (pamphlet), PCAIC: Cárdenas, Tabasco, 1979.

Map 7.1 Land Use on Ejido C-29, 1965

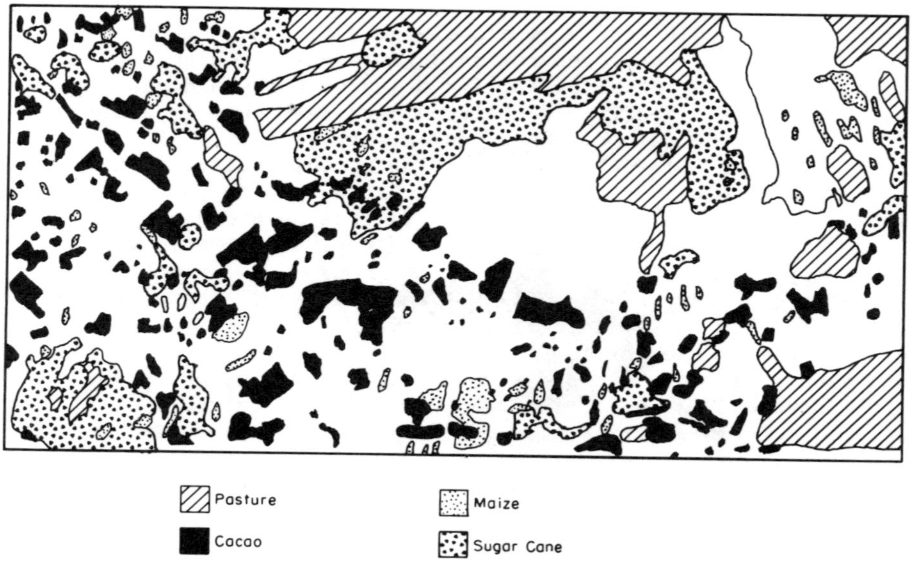

Source: Mexico, SARH, Promotora del Complejo Agroindustrial de la Chontalpa, "Uso de Suelo C-29, 1965."

Map 7.2 Land Use on Ejido C-29, 1980

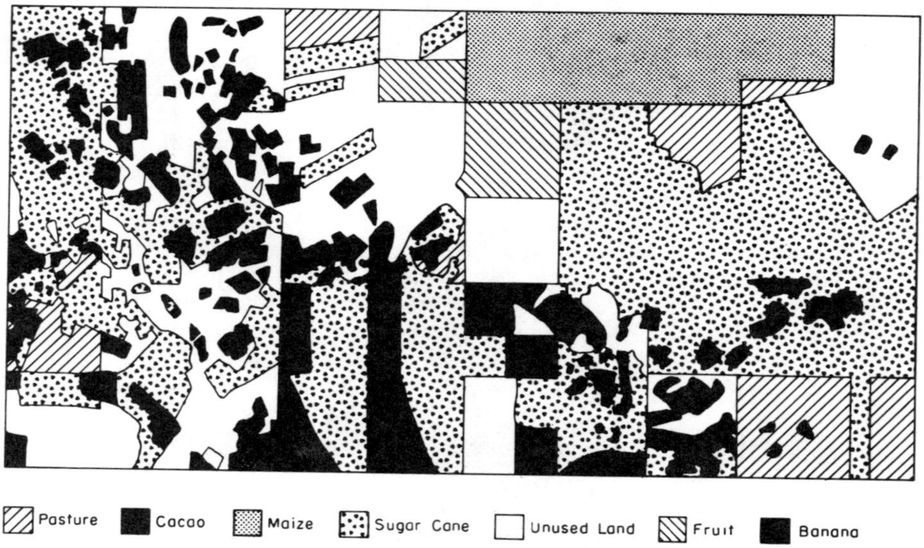

Source: Mexico, SARH, Promotora de Complejo Agroindustrial de la Chontalpa, "Uso de Suelo, C-29, 1980."

570 women; 357 of these were school-age children. Literacy rates for the parental generation were relatively high, although the average schooling for heads of household was only two years. Nearly all children aged 9-16 were in school; some older students had reached university level.

Twenty-nine of the families living in the poblado of C-29 were interviewed by the Sociodemographic Survey in January 1980.* All the collective families in that sample were young, most with school-age children, and the dependency ratio was thus very high.

Most ejidatarios were original inhabitants of the area, whose land was expropriated for the project. Of the original ejidatarios in the entire Plan, some 20 percent had emigrated and been replaced by landless workers; this estimate probably applied to C-29 as well. At least 15 ejidatarios there still owned outside properties, and hence worked only part-time on the ejido, compared with 20-25 percent Plan-wide. Chart 7.2 summarizes the main demographic and employment characteristics of collective cacao farm households.

COLLECTIVE CACAO PRODUCTION

Table 7.1 shows the 1980 agricultural program planned for C-29. Cacao provided half the value of total agricultural production (itself 97 percent of all production value), while accounting for only a quarter of ejidal costs and capital. Three quarters of ejidal profits came from cacao, a quarter from cane.

Most of C-29's collective cacao had been planted in old maize, banana, and cane fields. There was no credit available for cacao until 1972, but between 1972 and 1975, $12 million was borrowed to meet variable costs for cacao and $4.0 million for investment. In 1980, however, only C-40, C-26, and C-17 used cacao credit. This was due both to greater capacity for self-financing and to resentment of the tight production controls required by credit agencies. In 1979, the collective sector of C-29 had 179,993 mature cacao trees on approximately 280-300 hectares, and 9,749 trees on 56 hectares of newly established cacao.

---

*Approximately ten percent of poblado families were randomly chosen, but they were unrepresentative of the ejidal population as a whole, as the sample did not include the "libres" who live mostly outside the poblado center.

Chart 7.2 Demographic and Employment Characteristics of Collective Cacao Farm Households

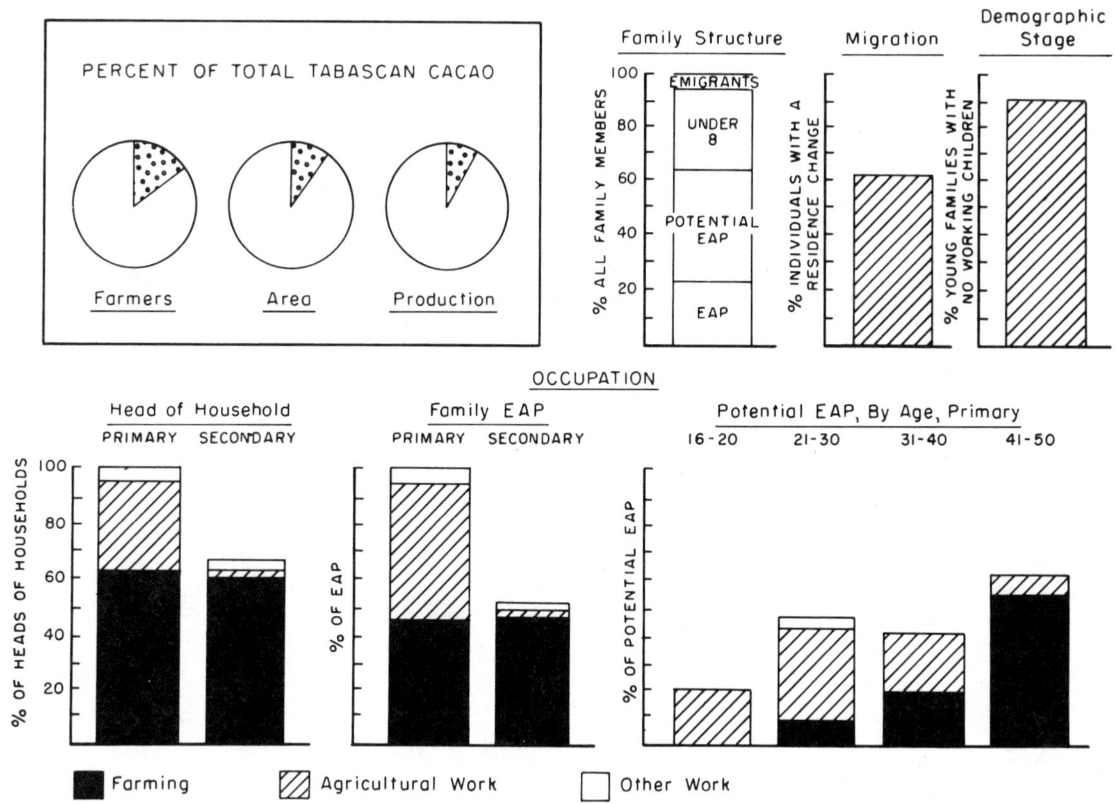

Table 7.1 1980 Agricultural Program for Ejido C-29

	Sorghum	Cacao	New Cacao	Sugar Cane	Livestock[a]	Total
Land area	100	579	21	1,166	589	2,455
Percent area	4.1	23.6	.9	47.5	24.0	100.0
Agricultural capital (%)	3	26	1	71	--	$19,825.9
Value of production (%)	2	49.5	--	45.6	3	100.0
Cost of production (%)	2.9	24.7	--	67.5	5	100.0
Expenses of costs (%)	9	67	--	43	48	--
Number of man-days	546	36,773	799	65,692	10,886.7	114,696
Percent man-days	.5	32.1	.7	57.3	9.5	100.0
Percent profits	1.0	72.5	--	24.5	2	100.0
Net income (including labor payments)	$173,432	$19,151,516	--	$11,413,064	--	--
Value per man-day worked	$317.64	$520.80	--	$173.74	$37.00	--

[a]The program for livestock had the following characteristics: number of head, 1,660; cows:heifers, 1:.72; culling rate, 1:8.77; bulls:cows, 1:20; number of steers sold:herd, 1:23; cows:calves, 1:.8; number of animals sold:herd, 1:12.6.

Source: Promotora del Complejo Agroindustrial de la Chontalpa, "Programa 1980," 1980, Ejido C-29.

Cacao Technology

The ejido did not apply insecticides regularly and had only recently begun a systematic program for pruning. Very low levels of fertilizer and fungicide were used. The cacao técnico complained of a 20 percent loss of pods on the tree from rodents, although cacao bean storage losses were low. C-29 growers claimed that yields had responded vigorously to the limited fertilization; 20-30-10 was used for flowering and 54-27-54 (scantily) for the soils. Weeding was done three times a year and funigation twice, but during 1980 the collective was far behind the recommended schedule on these activities, due to the preoccupation with cane production.

Shade used in the Chontalpa was predominantly moté. There was growing interest in replacing *Erythrina valutina*, which was subject to many pests and diseases, with *Erythrina glauca*. Chipilín was used only for temporary shade. Most planting was on a 4 × 4 grid, but some areas had triangular or other spacing.

The Cacao Plan technical adviser attributed the poor technology to lack of funds, particularly the distribution of profits rather than their re-investment, and also to competition with cane for labor. The ejidatarios said they were not particularly interested in increasing inputs for cacao, but rather in stabilizing inputs to keep cash investment relatively low.

Production data were not very reliable, but records showed that in 1978-79, a very good cacao year, the ejido averaged over a kilo per tree, 657 kilos per hectare. Average state yield was 739 kilos per hectare. By historical comparison, we find that overall Chontalpa Plan cacao productivity rose from 150 kilos per hectare in 1965 to 320 in 1970, dropped to around 190 kilos 1971-73, and rose to 370 kilos in 1975 (Promotora 1979). Due to the heavy rains in 1980-81, there was a 30 percent decline in cacao yields from the 1979-80 period.

Only about 60-70 percent of C-29 cacao was sold through the Collective Ejidal Cacao Association #23. Producers preferred to send cacao to associations, such as #23 and #2, where they were paid in cash rather than by check. The multiple outlets made data on monthly production difficult to collect.

In 1975, the first collective fermenters were installed in neighboring C-28, but C-29 did not ferment its cacao and sold all its beans sun-dried. This was ostensibly due to lack of warehouse space, labor shortage, and uncertainty about príce dífferentials between fermented and nonfermented beans. There was talk about the Union of Ejídos erecting another fermenting plant on a collective ejído, but details had not been worked out.

Production Costs

In 1978-79, the ejido earned nearly $11 million from cacao. Of this, a third went to labor costs--half of which was paid to ejidatarios with rights. An additional 38 percent went to pay profits to the ejidatarios with rights, proportional to the number of days worked by each. A fifth of the proceeds was reserved for a capital fund which paid labor costs for the following year and provided investment capital.

Chart 7.3 shows monthly costs of cacao production on C-29; more detail may be found in Scherr (1983b, Appendix Table E-5). Of cash expenses 1978-79, a fifth was for nonlabor inputs, a third for harvest labor, and nearly half for nonharvest labor. Cash expenses were highest in February, September, October, and December (over $500,000), and lowest April-July. Nonlabor costs were greatest in February and March during fertilization and in October for chemical spraying. Overall, in 1979, C-29 had higher costs and income than any other collective ejído.

The proportion of labor in total costs was high, even considering the low level of inputs. Even if input costs were to triple, the labor share would decline only to 67 percent. This compares to a labor share in cane production of 43 percent, for sorghum of nine percent, and for cattle operations (not investment) of 86 percent. Corresponding cost estimates for 1972 showed salaries accounting for 99.5 percent of costs.

ORGANIZATION OF LABOR ON EJIDO C-29

As the population density of the Chontalpa rose with ejidal settlement, a chronic labor surplus developed. Early agricultural efforts largely failed, and the movement to cattle did not, after the first infrastructure investments, provide adequate employment. During the late 1970s, however, a number of factors led to a labor *shortage*: increasing capitalization, increased land in use, the predominance of labor-intensive cane and cacao production, and an active off-farm labor market. These factors were particularly marked on C-29, and acted as a major constraint on ejidal production.

Historical Labor Use

In 1972, C-29 ejidatarios worked only 18 man-days per month and 216 days per year. At that time, the value of cacao production per man-day was only $18.60 vs. $184 in sugar cane (Barkin et al. 1975, p. 143). Labor use that year was only

Chart 7.3 Cacao Production Costs on Ejido C-29, 1979

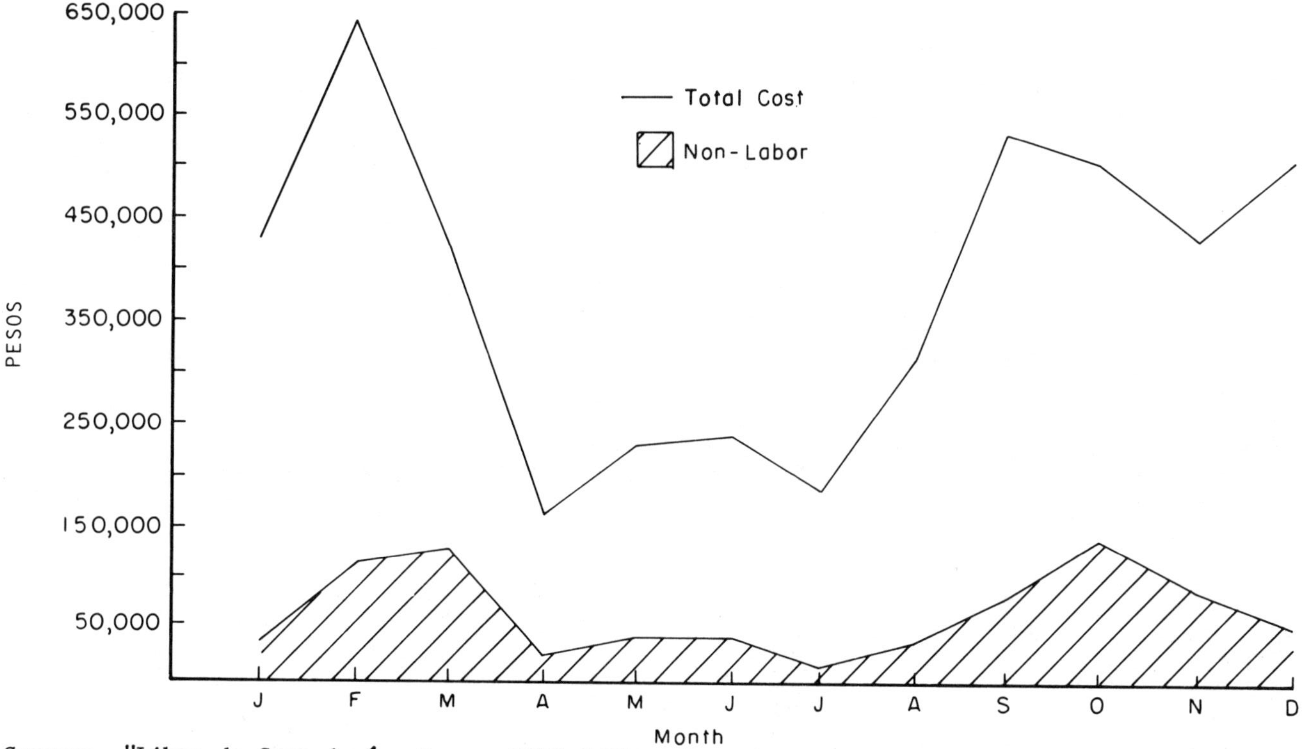

Source: "Libro de Contaduría, Cacao, 1979," Plan Chontalpa, Ejido C-29, Tabasco, Mexico.

one quarter the level projected for 1980, although there were 121 ejidatarios then, only a quarter fewer than at present (Barbosa 1974, p. 148). Nevertheless, C-29 had the highest level of employment in the Chontalpa Plan at that time.*

In 1976, two thirds of the labor for collective production was provided by ejidatarios, 28 percent by libres, and six percent by women. Labor requirements varied from 4,700 to 10,000 man-days per month. Between 1975 and 1980, the amount of labor generated by the ejido rose 29 percent; labor demand in the peak month of February rose 40 percent (Barkin 1976).

In 1976, there were 14 types of workers in the Chontalpa,† which included (Barkin 1976):

1. directors (political, administrative, foremen);
2. representatives of support institutions (e.g., credit);
3. specialists such as machine operators, cattle managers;
4. ejidatarios working year-round in stable jobs such as storehouse caretaker or cowboy, for wages and a share of profits;
5. ejidatarios doing jornal work on their own ejidos, under the supervision of zone chiefs, for wages and a share of profits;
6. sons of ejidatarios who were over 16 and were paid an established wage, but no profits, and who work on the family plot as well;
7. nonejidal jornaleros used as substitutes, who were paid wages and sometimes a share of profits (especially by the female ejidatarios);
8. ejidatarios working as jornaleros outside their own ejido, who got priority over libres, but no profit share;

---

*Average employment at this time was 11 days per month and 140 per year (Barkin et al. 1975, p. 148).

†Prior to collectivization, there were other types of work relations, such as the *tequio* and *convite*. Because of the earlier scarcity of labor and difficult access to peons, even largeholders were involved in shared work. There has been a steady decline in the use of tequio and convite, and a subsequent rise in importance of the ties of compadrazgo involving mutual work obligations and favors, usually for the private plots.

9. "enganchadores" who arranged for groups of jornaleros to work, especially in the cane harvest;
10. "libres" who may or may not live permanently on the ejido, who receive wages but no profits or social security;
11. jornaleros (usually ejidatarios or their sons) contracted by institutions, especially for machinery operation, who get no profit share, but did receive room, transport, tools, and a wage much higher than minimum;
12. jornaleros contracted by individual ejidatarios for work on the private plots;
13. women working in the women's collective plot, on collective work operations, or at the banana packing plant;
14. technical personnel paid by the ejido, but usually contracted by institutions (such as teachers).

Labor Patterns in 1979-80

On C-29 in 1979, the labor force included the 161 ejidatarios working full-time (or hiring replacement labor), perhaps 100 sons of ejidatarios without rights, 40-70 resident libres (of whom 30 worked permanently in cacao, receiving housing and social security as well as a wage), and 10-20 "eventuales" living near the ejido who come regularly for work. This local pool satisfied the ejido's labor needs six months of the year; in December-March, August, and October, large numbers of outside workers were needed.

Total labor used by ejidatarios with rights in 1979 was 38,572 man-days. Monthly labor use ranged from a low of 2,656 in December to a high of 3,692 in October (see Chart 7.4). The average number of days worked per year by ejidatarios with rights was 241; of these almost two thirds were in cacao, fulfilling half the cacao labor requirements. Cacao accounted for 40 percent of all labor used.*

Employment for most ejidatarios varied through the year. First cacao harvest, then pruning, occupied January-March at which time the ejido employed outsiders for the cane harvest.

---

*Note that labor use data for 1979 for all 22 ejidos showed at most 1,067,250 man-days worked by ejidatarios with rights, and at least 3,769,480 worked by others--over three forths nonejidatarios.

Chart 7.4 Monthly Labor Use on Ejido C-29, by Enterprise (January-December 1979)

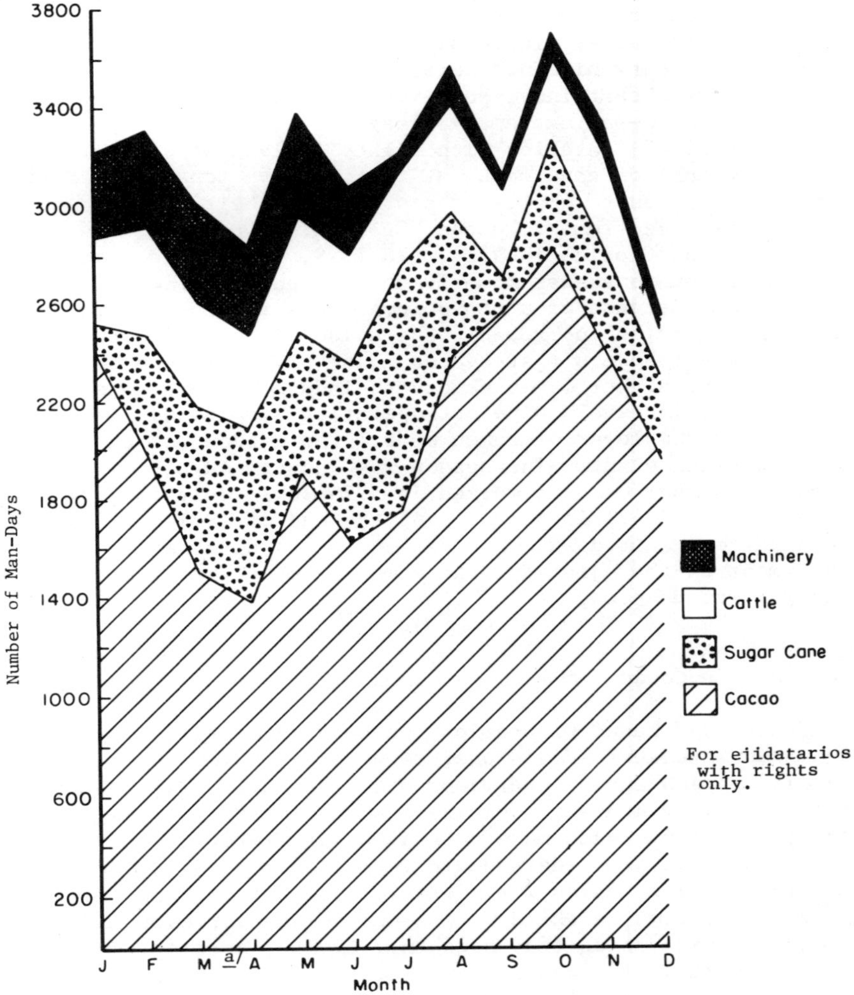

[a]Data for 34 ejidatarios are missing for the month of March.

Source: Ejido C-29, Plan Chontalpa, "Libro de Aportación, 1979.

When the latter returned to their own subsistence farms for the maize planting, ejidatarios completed the rest of the cane harvest, which lasted through June. Cacao work at this time was often done by outsiders. Beginning in August, there was a sharp and steady increase in cacao activity by ejidatarios and by many hired workers.

Ejidatarios certainly, and even local libres, were reluctant to cut cane--an extremely disagreeable job--despite a premium wage. Ejidatarios with rights in 1980 were required to cut 200 tons of cane each, at four to five tons per jornal. They were allowed to employ others to meet their quota. Canecutters came from Michoacán, Guatemala, Guerrero, and from Jalpa and Cupilco locally. In 1979, 100 people came from Michoacán to assist some 70 socios and a few local jornaleros. Because of a ten percent loss in cane production due to heavy rains in 1980-81, the ejido did not hire contracted cane laborers that season. The price of cane was such that the higher wages needed to induce workers to cut cane would effectively wipe out operating surpluses.

The main employer of women on all the ejidos was the banana packing plant. The season was January-August, with a maximum number of days worked in 1975-76 in July (1,553) and a minimum (242) in August. At the end of 1981, the packing plant was shut down. Whether there was alternative employment on the ejido for these women (mostly single mothers) was unknown.

Labor Use in Cacao

Until 1980 cacao plantations on C-29 were typically organized into seven divisions, each with a chief chosen by the ejidatarios and a supervisor over them all. The chief daily chose the men who would work with him for a given chore; at harvest each managed a group of 12-20.

The number of man-days dedicated to cacao, and the role of ejidatarios in cacao work, are illustrated in Chart 7.5. Of total man-days, a third were spent on the harvest, nine percent in processing (pod-breaking, drying, sorting, carrying), a quarter in pruning, 14 percent in weeding, and six percent each in administration and chemical spraying. Because there was no month when ejidatarios with rights were unemployed, there was little internal pressure to add activities such as improved pruning, because the labor would have had to come from hired workers.

The lack of individual responsibility for the condition of the plantations led to disappointing production results, even with the addition of new inputs. Cacao was not mechanized, so cultivation was dependent on careful hand work; the collective system was not tailored to this type of production system.

Chart 7.5 Labor Use in Cacao on Ejido C-29, by Type of Worker, 1979

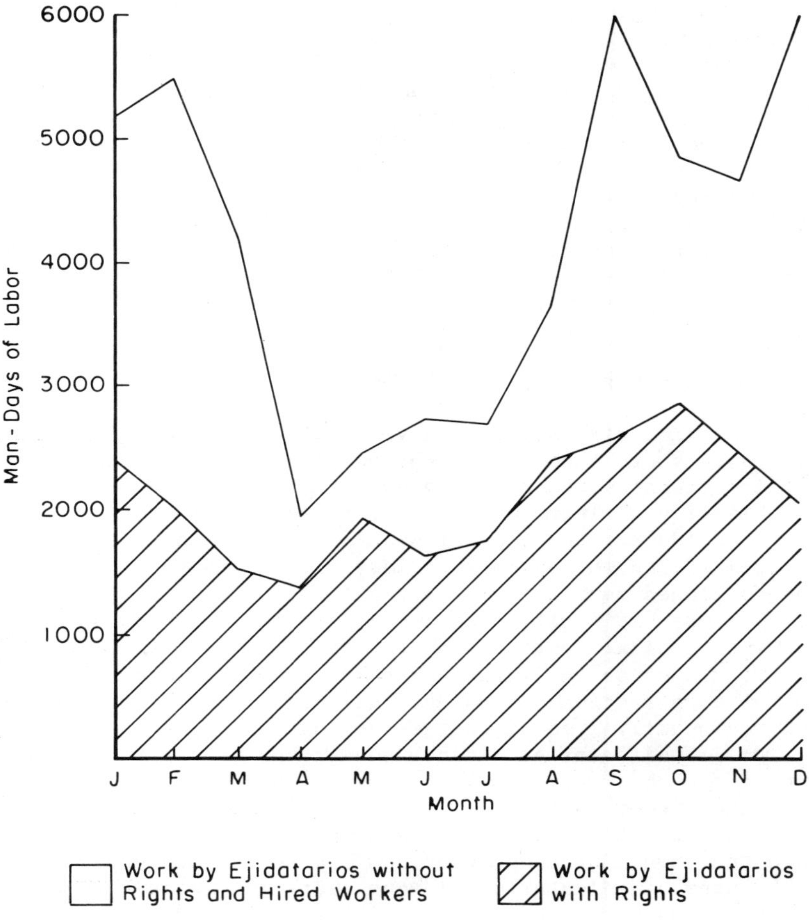

Source: Ejido C-29, Plan Chontalpa, "Libros de Contaduría, Cacao, 1979."

In 1980 the system was changed experimentally to one used earlier in C-09 and C-10, in which each ejidatario was given responsibility for a particular area in cacao. Plantations were divided into areas of 4,800 trees between four ejidatarios with rights, with each responsible for 1,200 trees. They did not receive a jornal, but were paid for the harvest. The ejido provided urea (charged at the harvest); the other inputs farmers bought from their own income. Farmers could hire whomever they wished to help with the cacao. A chief cacao manager checked the plantations to ensure that production was going forward as it should. If certain technical requirements were not met, the farmer would be relieved of his plot.

According to the agreement with the Secretaría de Reforma Agraria, whose job it was to assure compliance with national regulations governing ejidal organization, the farmers had to rotate plots every 18 months. Early results from the new program were promising.

Off-Farm Employment

Ejidatarios with rights did very little work outside the ejido, even where such work might be well paid (with the exception of those who still owned agricultural land outside the collective). There were strict rules against work outside the ejido for them; permission from the ejidal assembly was required and the ejidatario lost all rights to profits. If he continued such activity for a longer period of time, he could lose his position as an ejidatario with rights. There were similar restrictions for the resident "libres." In 1979, six or seven libres had taken leave to work full-time for construction companies; they had to pay house rent--which was usually free of charge--during that time.

Wives of ejidatarios did not seem to be as active either in outside work or home work for pay as they were in other types of producer families. The hunting, fishing, and gathering activities prominent as supplemental income sources before the Chontalpa Plan began had slackened considerably.

The sons of ejidatarios did seek work in a wide variety of occupations but were far less likely to work as jornaleros in agriculture outside the ejido. The Benito Juárez sugar cane mill began to hire many sons of C-29 ejidatarios around 1976-78. They earned $165 per day as a part-time libre, more if they were unionized and had permanent work. Others were employed in the technical education centers within the collective ejidos, a few in ejidal administration. The rest were employed on private ejidal plots.

Between 1978 and 1980 many sons of ejidatarios (one estimate was about half, or 100 individuals) turned to work in petroleum boom-related construction. In 1979, of the 47 economically active individuals in C-29 surveyed by the SST, only two worked outside agriculture--one in tire-making, the other in house construction. Only four individuals had worked outside agriculture during one of the years surveyed, and that was in 1965.

Earnings and Income

Sources of earnings for C-29 residents included wages for labor, profits for ejidatarios, remaining indemnification payments, off-ejido work activities and properties, and sales from private parcel production sales. Wages in 1979 were $70-75 per day, with pay for piece work of $25 per sack of broken pods and $1 per tree weeded. Pay rose to $92 per day in 1980. This was the same for all workers (except enterprise managers) for a day lasting from 7:00 A.M.-12:00/12:30 P.M. In addition, some resident libres received Social Security benefits (as did all ejidatarios), plus a house (worth at minimum $5,000), and the right to a maize plot (estimated worth $50,000).

Imported laborers earned $120 daily for cane-cutting, as well as housing and a maize ration. If there was no work on a given day, the latter still received $50 and the maize ration. (Alternatively, they may receive a straight $50 per ton cut.)

Almost all the indemnification money had been used up or invested by 1980, but the properties outside the ejidos probably brought in more money than the owners earned as ejidatarios. Incomes from private ejidal plots ranged widely. For the plot with only maize and some home vegetables, the total value of production was only a few thousand pesos. For those ejidal families lucky enough to have land appropriate for cacao and whose trees were in full production, additional family income could amount to $20-30,000 pesos.

Until 1976, only cacao production had shown any profits project-wide. In 1979 it was certainly the best earner in C-29, as $5,256,000 were distributed as profits. Cattle operations also distributed profits, but much less. Each product was considered a separate "line," and those in charge could decide to distribute profits one to four times a year. Barkin wrote in 1976 of the importance of profit distribution:

> *The socios are always anxious for the distribution of profits; they await it so desperately that on various occasions, the Fideicomiso has made the distribution without there having been any profits,*

charging the sum to the standing debt of the ejido [Barkin 1976].

The total level of profit distribution throughout the Chontalpa ejidos was $38 million for 1978, $80 million for 1979, and $118 million for 1980. On C-29, in 1980, $3 million of profits were divided among 162 ejidatarios. Note that of the ejido's $4 million cattle debt, half had been paid off, and some forgiven by the government, leaving C-29 in a fairly enviable financial position as of the early 1980s.

In 1980 the average Chontalpa Plan family earned $54,000 (Promotora 1979)--$27-28,000 per worker (personal interview, director of Promotora). The directors of C-29 estimated that the average earnings for an ejidatario with rights were $50-60,000 per year in profits and $25,000 in labor--giving an income of $75-80,000--roughly comparable to many family farm households. Incomes on C-29 were probably higher than in any other collective ejido.

PRIVATE PLOT PRODUCTION STRATEGIES

Although they belonged to a collective in which many strongly believed, the ejidatarios operated also as independent families. That independent life was centered on the use of the private plot.

The plots originally emphasized subsistence crops such as maize and bananas; others produced some cane, coconut, cacao, cattle, and small livestock. But in 1980, according to the directors of C-29, between 40 and 80 percent of the ejidal private plots had cacao. From no more than thirty hectares in 1975, there were almost 200 hectares in 1980.

The cacao extension workers estimated that the age distribution of private ejidal cacao plantations was similar to that of the Chontalpa project as a whole: 30 percent of trees in production; 40 percent a year old; 30 percent 1-3 years old. In general, yields on private plots were higher than on collective plots. Many families used fertilizers bought from FERTIMEX and purchased insecticides from private stores.

The ejido provided credit for cacao produced on personal plots; four to five ejidatarios had to group together to receive it. But in 1980 very few used such credit.

The family had the following objectives:

1. Maximizing returns to family labor from all sources;
2. Maximizing use of the private plot land;

3. Maintaining the position of official ejidatario in the collective;

4. Maximizing protection for the family from economic swings in the collective.

There was a difference of opinion on whether the private plots competed with collective labor needs. The answer depended on how much access to family labor the farmer had. Under most conditions, the afternoon and a few weekends were sufficient for an ejidatario and one family member to complete all the necessary activities on the private plot. However, as of 1980, most of this cacao was newly planted and hence did not require harvest labor. Under the present collective schedule, the private cacao harvest would conflict with both peak collective cacao harvest and peak cane harvest.

Costs of production and organization were very similar to the minifundista family farms. (The corresponding "off-farm activity" in this case was collective farming.) Regardless of family interest in increased chemical input applications, the cash constraint was significant. Farm management was likely to favor a high ratio of labor used relative to cash inputs. The need for cash on private cacao plots added to the unwillingness of farmers to re-invest profits in the collective plantations. They preferred to receive the remanente to invest in cash inputs on their own family plantation.

BASIC STRATEGIES OF PRODUCTION ON COLLECTIVE CACAO FARMS

Decision making for the collective ejidos of the Chontalpa required monumental trade-offs in order to work out economically and politically. There were five basic objectives:

1. Meet production goals for politically determined products;

2. Maximize land utilization on the collectives;

3. Pay back initial debt of the ejidos;

4. Maximize returns to the ejidatarios;

5. Maximize employment for ejidatario families.

These criteria rarely pointed to the same pattern of resource allocation. Hence there was continual institutional conflict between the body of ejidatarios, their directives, the Union of Ejidos, the Chontalpa Promotora, the Banco de Crédito Rural, and the national SARH over which activities would be promoted and how resources were to be distributed between wages, profits, investments, production costs, and social services.

Politics in Production Goals

As the history of the Chontalpa Plan land use earlier in this chapter showed, there had been continuous change in product focus. Decision making about investments and land use had been almost completely bureaucratic, largely out of Mexico City. The huge investment made in the Plan Chontalpa could only be justified politically if the collective farms made a significant impact on national agricultural shortages. Thus the initial agricultural emphasis changed to livestock production and then, despite earlier failures in this area, was re-oriented to stress basic foods production.

Optimizing Land Use on the Collectives

Again, the enormous cost of the project and its supervision by national leadership put pressure on the collective ejidos to intensify land use and maximize production as much as possible. The collective nature of the experiment, and the fact that most of its chief political backers were fond of collectivization, meant that these pressures tended to sacrifice private plot production for collective production when there was a trade-off to be made. Because production goals were more visible and compelling than financial goals, there was a tendency for official técnicos and directors to push for high production levels at the expense of profits. Due to increased ejidatario participation in the late 1970s, however, this problem was diminishing.

Paying Back the Initial Debt

Government and banking representatives played an important role in organization and production decisions over the years. The Chontalpa ejidatarios were limited in their independent decision making because of the large federal debt which they owed for ejidal development--land clearing, drainage, other infrastructure, cattle development. Cane profits and part of recent cacao profits were generally directed to repaying past debt. Most cacao profits had gone to internal financing of cacao expansion. Earlier losses of the basic grains projects were absorbed by cacao and cane operations.

Many of the basic development loans were forgiven the ejidatarios as being well beyond their financial capabilities. But the ejidos had come quite a way toward paying off their cattle debt. If this can continue several more years, the credit crunch so severe in the early 1970s should ease. It will be interesting to see whether the national banking institutions will indeed permit the ejidos to become financially independent, as this will significantly reduce outside influence on ejidal production and investment decision making.

Maximizing Returns to the Ejidatarios

Ejidatarios sometimes complained that they did not feel like owners or farmers any more; that effective decision making had been wrested from them by the many institutions involved in the Chontalpa Project, and yet they were saddled with debt and most of the risk. There was a substantial basis for this allegation, and the resulting "worker" mentality of many ejidatarios seemed in large part justified. This attitude hurt productivity and the long-run capitalization of the project.

Ejidatarios with rights expected a return comparable at least to that which they would have had as an ejidatario with rights on an individual plot ejido. As we have seen in other case studies (with only one exception), such returns were much higher than $92 per day. Indeed, they ranged from $90 to $332 per day. Demands for remanentes were thus not unreasonable, particularly when we consider that the per capita land base for a Chontalpa ejidatario was much larger than his neighbors'. With the remanentes, the average income per day worked in cacao by an ejidatario with rights rose to $200. Ejidatarios were unwilling to sacrifice these profits to investment, the returns from which often went to others or could be absconded by a dishonest leader.

Maximizing Ejidal Employment

As we have seen, there was in the late 1970s and early 1980s no employment problem whatsoever for the ejidatarios with rights, who had first priority on jobs. But they were far outnumbered by their offspring, whose access to work was less secure. On C-29, a large number of them could be employed almost year-round, but the wage was so much lower than the going wage in the surrounding area that to achieve a comparable income they needed to work many more man-days.

This ejido had certainly fulfilled the goal of producing jobs. Using 1980 project figures (which did not include labor used on private plots, nor employment of resident nonejidatarios in Union industries), the number of man-equivalents (meq) per hectare for the entire 5,000 hectares was 12.2. If we estimate that indicator based on land in use (subtracting 1,300 hectares of acahual), the figure drops to 9.03 meq/ha; subtracting all land not in commercial production (a total of 3,320 hectares), it drops to 8.10. Considering the importance of cattle activities, this was a very high rate of labor intensity, though not as high as the minifundia cacao farms just outside the project limits, which of course had lower capitalization and lower incomes.

The ejido had been transformed into a large-scale employer, quite incongruent with the standards for family-based ejidos, which were based on the concept of "land to him who works it." There had been an attempt, in the case of ejidatarios without rights and some favored libres, to confer secure employment and extra benefits in order to avoid the exploitation they associated with capitalist employment. To the dissatisfaction of many with nominal rights, the line between ejidatario and nonejidatario was blurring. The collective ejidos of the Chontalpa already operated essentially on the basis of corporate capitalism in their dealings with the rest of the economy, and the importance of nonejidal workers was in direct contradiction to the precepts of ejidal development.

This was a critical issue, because labor was the principal constraint to increased production on C-29. Resolution of the labor constraint could only come from further incorporation of nonejidatarios. If the Chontalpa ejidos are intended to be government projects maximizing economic and social benefits for the agricultural population, then its employment policies will be different than if it is expected to maximize incomes to ejidatarios with land rights on the ejidos.

Increasing Productivity

Experiments with new cacao technology were intended to reduce some of the above frictions by increasing productivity. A major new feature was the cross between Guayaquil and Criollo to yield a variety which began to produce at two years and had very high experimental yields. The Union talked of aerial spraying for cacao, similar to that frequently used in Tabasco for large banana plantations, but such a technology required compact squares of 100 hectares of trees to be profitable, while most Chontalpa cacao was disperse.

There was substantial interest in expanding cattle production, because a relatively small increase in employment could yield significantly higher profits. But this was limited by a shortage of feed, especially in the dry season when the ejido had to sell animals for this reason. The ejido planned to increase the area in improved pastures (African Stargrass, Alemán, Elephant Grass) from 380 to 880 hectares. There was a little supplemental feeding of animals with grains and cane by-products. Earlier experimentation with silage had failed. The technical constraints for increasing the feed base were significant.

The collectives certainly had some potential advantages for implementing new technologies to improve productivity. But the organizational and employment problems needed to be dealt with first. The technology already available was far superior to that which was in use; indeed, family farm cacao technology was better than that used on the collectives.

RESPONSE OF COLLECTIVE CACAO FARMERS TO ECONOMIC CHANGE

The economic and social patterns of the Chontalpa ejidos had changed drastically in the 1970s. Though mid-decade studies indicated little improvement in real living standards among collective ejidatarios, by 1981 there was certainly greater contentment and prosperity.

Collective farm response to changes in prices and wages, and the general reaction to the petroleum boom, were far different from those of other groups. The collectives were extremely responsive to changes in product prices--once basic production requirements set by SARH or Mexico City had been met. If cacao prices should rise, there would be a greater shift in resources to cacao production, reducing levels of cane production to the minimum politically required. All potential producing land was being planted to cacao.

If prices decline, however, the ejido will be under tremendous pressure to improve cacao productivity, which may not be possible without a several year period of declining profit distribution in favor of investment. Whether this is possible without coercion during a time of declining total incomes is debatable. There may be a strong temptation to transfer resources instead to cattle-raising and pasture care, particularly if the labor scarcity continues. Since the ejidatarios did little work off the collective, they would not be much affected by the decline in general cacao labor demand which would occur if prices decline.

Immediate investment plans indicated a balanced interest in diversification. Cacao area was to increase by about 400 hectares. But much unused land was also to be cleared for pasture; land along the drains (its use had formerly been restricted) was to be used for subsistence maize production. The ejido planned to increase its investments in cooperative agroindustry within the Union of Ejidos. Between 1981 and 1984, C-29 planned to add a hog enterprise as well.

With the large increase in cacao area, livestock enterprises, and subsistence production, there will be a sharply increased demand for labor on the ejido by the mid-1980s. The ejido hoped to mechanize more, particularly in cacao spraying and fertilization. But the demand for nonejidatario labor would certainly continue to rise. If the trend continues on other collectives, they promise to be among the principal employers in the region.

As such, they will be very sensitive to the going wage rate for agricultural and nonagricultural labor. And since it is unlikely that ejidatarios would submit to being paid less for a jornal than hired nonejidal companions, any increases in the general wage will inflate labor payments to ejidatarios as well.

The increased demand from the collectives will itself push wages higher; if this combines with significant employment by PEMEX drilling on ejidal land, the ejidal budget could be sorely beset by labor costs--in a manner somewhat similar to the hired labor based producers today. There would likely be very strong pressure to reduce sugar cane production; more intensive cacao technologies are unlikely to be successfully introduced in such a scenario.

The ejido's response to declining wages would also be similar to that of hired-labor-based producers--at least so long as hired nonejidatario labor was important in breaking production bottlenecks. If wages went quite low, ejidatarios without rights would probably demand greater employment on the ejidos, which would partially displace hired nonejidal workers.

ROLE OF COLLECTIVE AGRICULTURE IN A PETROLEUM ECONOMY

As of 1981, there were 15 oil wells situated on C-29 land; some 70 others were planned for some time in the future. With each of these, the ejido lost the use of three to four hectares around the well, as well as land cleared for road-building (even some cacao has been cut), and some land from clogged drains. Land tenure was secure, but indemnification payments (like those to all ejidatarios) would cover only land improvements, not the value of land lost to them.

But despite PEMEX employment of ejidatarios' sons, the self-contained internal economies of the collectives had been relatively sheltered from the petroleum boom's effects. The cost of living impact was marked, since nearly all the ejidatarios' commercial activity took place in Cunduacán and Cárdenas, two of the major oil towns. But many of the employment-related and other problems of the collective ejidos were as much a reflection of internal development dilemmas as of the petroleum boom.

The study of C-29 yields a number of conclusions regarding guidelines for tropical project development, the distribution of activities between private and collective jurisdiction, and the role of the collective in a petroleum economy.

First, as has been often, but cannot be too often stated of tropical agricultural projects: it is of fundamental importance to understand the environmental characteristics and constraints of the land before undertaking expensive, semi-permanent infrastructure investments. Some land use changes in the Chontalpa were due to political priorities; too many were due to unforeseen inadequacies of soil and water control conditions. Production still suffers from the early decisions taken from mistaken agronomic information. Precipitate action like the destruction of so many cacao plantations, which represented years of investment, was extremely unwise, particularly in a new project. Indeed, the record of large-scale mechanized clearing projects in the Mexican tropics is abysmal (Ewell 1984; Price and Hall 1983; Yates 1981).

Second, many activities can be undertaken by collectives more efficiently, productively, and equitably than by individual farmers. Large-scale infrastructure management and investment; marketing; input purchases; product processing; organization of schools, bus lines, and other local services are examples. Collectives provide easier access to credit and mechanisms for risk-sharing.

However, basic agricultural activities and decision making need to be decentralized and are much better left to individual farmers. Success in farming, more than in most other productive activities, is dependent on constant care, day to day flexibility, and sophisticated management of microenvironments. Organizational changes in C-29--for example, making individual ejidatarios responsible for specific cacao plots--should serve to strengthen, not weaken, the community-based production system.

Yates noted that most other theoretical advantages of collectives--access to mechanization and the consequent replacement of grazing land (for work animals) by crops; higher quality technical services, advantages of scale for modern production methods--have not worked out in practice, particularly in tropical areas where the large-scale technologies are neither well-developed nor superior to smallholder technologies. Monoculture increases the hazards of nature; centralized decision making increases the hazards of bad management (Yates 1981, pp. 168-70).

Third, collectives play a peculiar role in the rural labor market, as seen in the case of C-29 in Tabasco. In

Mexico, ejidatarios on individually organized ejidos were not allowed to hire workers full-time, yet the collective had been forced to do so and would continue to expand such employment. There was an urgent need for a new code relating to hired labor use on all ejidos, to increase total production and encourage increased employment of landless workers. Some serious political problems may need to be resolved about the status of these workers, their rights, and their rewards.

Fourth, collective decision making by hundreds of workers is not easy to carry out; it is particularly unlikely to be successful when a project is subject to heavy outside political pressures. The common alternative in large projects is development of a hierarchical and authoritarian decision-making structure, which can only be called "collective" in the most superficial terms. Management is then dominant and ejidatarios dependent, in a way reminiscent of hacendado/peon relations in the last century (Yates 1981, p. 172). The risks of large projects, particularly ones subject to intensive political scrutiny and control, and subject to only minimal farmer decision-making input, should not be borne by the farmers. Such projects should clearly be seen and managed as state farms; otherwise, farmers should be given real decision-making powers.

Fifth, under conditions of relative labor scarcity, such as take place during a petroleum boom or in petroleum-producing regions, the collective mode of production does not appear to have many advantages over smallholder production systems, if the latter have access to well-organized input and output marketing services. Labor constraints, high labor and other costs, and inflexible management are particular constraints for collective production.

The principal advantages, as seen in the case of C-29 in Tabasco, are relative insulation from the outside economy and direct access to public resources and subsidies--which could be used to offset petroleum-related problems. But large-scale development projects tend to be excessively capital-intensive and extremely costly, with most of the expenditures accruing to construction contractors and relatively well-paid professional and service staff, rather than to farmers and farmworkers. In a petroleum economy, where precisely these services and materials are in extremely high demand, such projects are at a further disadvantage.

8

Cacao and the Rural Labor Market in Tabasco

The composition of the labor force in Tabasco cacao production underwent fundamental changes as a result of both the petroleum boom and long-term trends within the cacao sector itself. This chapter discusses the role in state cacao employment of farm family members from different types of households, analyzes the statewide supply and demand for landless agricultural workers as they relate to the cacao sector, and examines the impact of petroleum development on the structure of the rural labor market in Tabasco.

THE COMPOSITION OF THE CACAO LABOR FORCE

Chart 8.1 illustrates the total number of workers in the cacao labor force in 1979-80 estimated from the SST. Including secondary occupations, the labor force numbered about 25,000 workers--some 16-20 percent of the state's EAP in agriculture and 35-45 percent of the EAP in the six principal cacao-growing counties. Of this total, about 52 percent were farmers (as either the primary or secondary occupation); 29 percent were farm family members (working either on their family's or others' farms); and only a fifth were workers from landless families. The number of landless is probably underestimated due to survey problems (see Appendix B), but the SST data certainly confirms a major decline in the number of landless cacao workers during the oil boom. Figures on historical employment suggest that landless workers, formerly in the cacao sector, were seeking employment elsewhere. Their surprisingly limited participation underscores the fact that Tabasco's heartland has remained predominantly smallholder, with rural proletarianization not yet advanced.

Chart 8.1 The Cacao Labor Force in Tabasco, 1979-80[a]

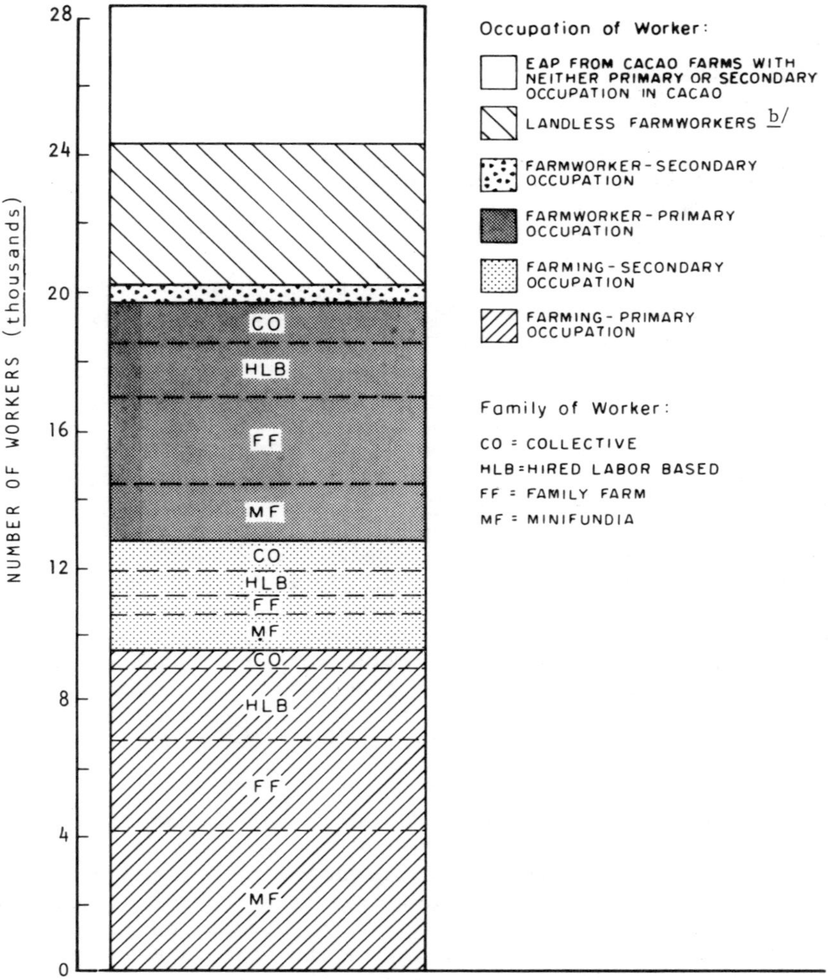

[a]The total EAP in agriculture in 1970 was 116,147. Thus, cacao work accounted for 20 percent of agricultural jobs. If only cacao municipios are included (Comalcalco, Cárdenas, Cunduacán, Jalpa, Huimanguillo, Paraíso), the agricultural EAP was 53,630, and cacao accounted for nearly half of all workers.

[b]This figure actually ranges between 3,700 and 4,900, depending upon SST sampling interpretation.

Source: COPRODET and CONAPO, Encuesta Sociodemográfica del Estado de Tabasco, Villahermosa, Tabasco, January 1980; extrapolations from Cacao Subsample.

Of the labor force from farm families, 39 percent came from minifundia, 32 percent from family-labor-based farms, 19 percent from hired-labor-based farms, and ten percent from collectives. Note particularly that though the estimated 12,649 farms* (including collective private plots) had fewer than 23,000 active workers, the *potential* EAP from these families was 64,886. Even if children 8-14 were excluded, there was obviously considerable labor available that could have been tapped had labor returns warranted.

Furthermore, a fourth of cacao farms had *no* worker whose primary occupation was in cacao farming. A *quarter* of the actual EAP on cacao farms worked outside agriculture as their primary occupation, usually in skilled trades or in commerce and services. This means that they were not counted in the agricultural EAP in the state Population Census.

These statistics clearly support our earlier impression that the family work force available for cacao production was extremely flexible. If half of each of these categories of individuals were to dedicate themselves to cacao farming, an additional 24,000 workers could be added to the available labor force--or almost two per farm in the state. This was equivalent to well over ten percent of the entire labor force in agriculture in Tabasco--enough for their entry and exit to make a major impact on wages and the tightness of the agricultural labor market.

Characteristics of Cacao Farm Workers

The SST provides us with a profile of cacao farmworkers, seen in Chart 8.2. (Summary statistics may be found in Scherr 1983b, Appendix Tables C-21, C-22, and C-23.) Cacao workers tended to be from larger families than average, with more members of working age. They were more likely to come from hired-labor-based and family-labor-based farm households, and less likely to be from minifundista families than was the "average" worker.

Nearly a third were heads of households--but these came almost entirely from collective and nonfarm households. The rest were children of household heads and, as would be expected, were young. About half of all cacao farmworkers were under 20 years of age.

There was a big difference in age distribution by household type. Nearly three quarters of those from minifundia

---

*This is based on extrapolation of the total number of farms in the sampling from 2/100. Cacao Association statistics showed 14,000 producers (not clearly defined). By 1984, their estimate had been changed to 16,000 producers.

Chart 8.2 Demographic and Employment Characteristics of Nonfarm Households

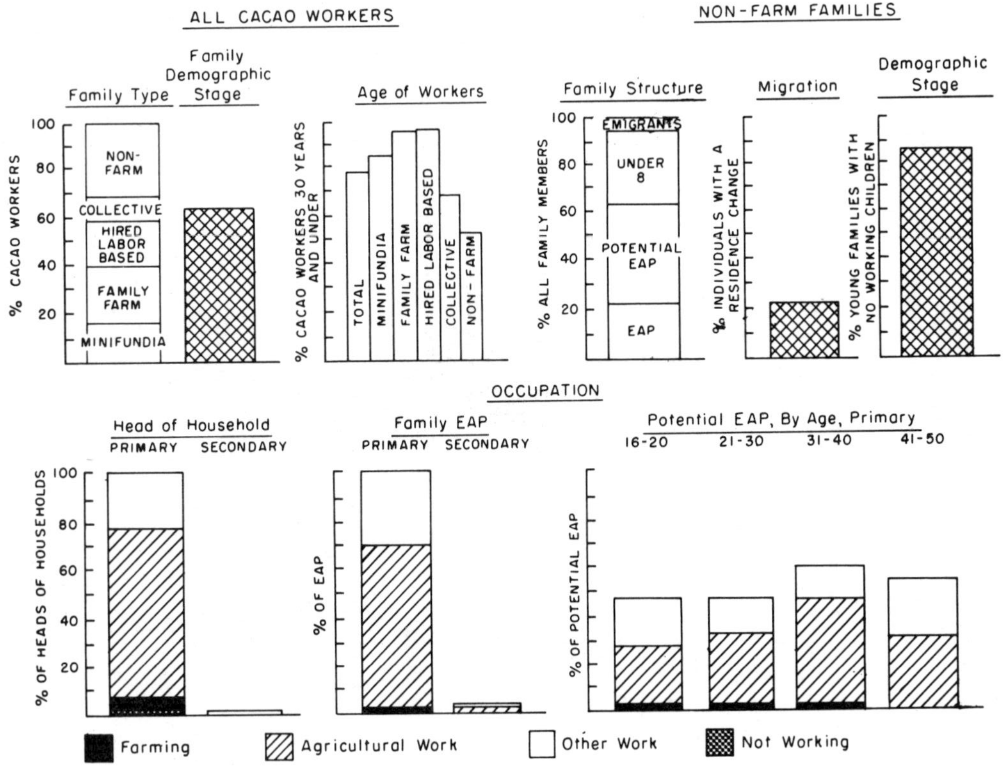

Source:

families were under 20, about 60 percent of those from family- and hired-labor-based farms. Only 23 percent of farmworkers on collective farms, and 19 percent in nonfarm families, were under 20. Only 2.6 percent of cacao workers were also cacao farm owners in 1979.

LANDLESS AGRICULTURAL WORKERS: FAMILY STRATEGIES AND CONDITIONS IN THE 1970s

Landless households in Tabasco in the 1970s derived family subsistence from multiple employment. The head of household had several sources of income, and/or his wife and children, including schoolchildren, participated in economic activities. According to the SST, families in the nonfarm group had two economically active members twice as frequently as the "average" family. There were fewer families with five or more active members (38 percent vs. 51 percent on average). They rarely had savings accounts, their only economic back-up being the raising of small livestock for sale when needed. They usually lived on other people's land.

The dependency ratio tended to be much higher than for the average cacao farm family. This was because the predominant demographic stages were the "young married" and "school-age child" stages. In general, the population was younger, as evidenced by the age distribution of household heads. Nearly three quarters of household heads were under 40, versus about a third on average.

For the landless population as a whole, cacao work and agricultural work in general had been declining in importance. This was particularly so for younger family members, whereas heads of households and workers over 30 tended to be involved in agriculture.

The SST shows that of the actual EAP in nonfarm families, three quarters worked for a patron and ten percent for the family. But only 57 percent were jornaleros and 12 percent were themselves employers. Less than three percent worked for PEMEX or a constructora. Although 70 percent did work in agriculture in 1979, large proportions worked in skilled, unskilled, and commercial sectors (15.5, 4.7, and 9.3 percent, respectively).

Of those who formerly worked in cacao plantations, but did so no longer, a fifth were no longer working at all. Of the rest, only 18 percent still worked in agriculture; over half had switched to employment in skilled trades (see charts in Chapter 3).

A career in a skilled trade was by far the most lucrative and stable of all the possibilities open to the landless worker. But under the apprentice system commonly used, the master had to be paid for the apprentices' training time.

The financial resources and time required to develop these skills put the landless at a disadvantage.

Employment opportunities for the landless were certainly available in a booming economy such as that of Tabasco. But this did not negate the extreme precariousness of much of the employment and their vulnerability to slight swings in economic growth, particularly of the construction industry, and changes in agricultural price levels. The lack of organization and minimal job protection meant that much of the burden of economic dislocation was borne by this group.*

Finding Work

The process of finding work was itself quite time-consuming. For agricultural activities, workers often grouped together and went from farm to farm seeking work. In other communities, farmers visited the neighboring ejido's main square announcing available work and offering jobs. Where an individual had the custom of working with a particular patron, the two devised a system for communicating when work is available or required. Sometimes, if the worker's family was in difficult economic straits, the patron offered extra work. In general, though, the practice of significant nonwage benefits or gifts for workers from long-time patrons was disappearing.

To find PEMEX or constructora work, workers listened for radio ads, watched for public and printed announcements, and visited the potential places of employment regularly to see if jobs were being filled. Very frequently, personal connections were required in order to be granted one of these jobs. Sometimes the job search involved repeated visits to Villahermosa or Cárdenas.

STPRM operated offices in Cunduacán, Villahermosa, and elsewhere. In Cunduacán, it provided ten drilling teams of 30 people each, plus about 450-500 other workers in May 1980. About 100 libres were also hired for hauling chemical materials. They received minimum wage for that category of work, and were organized through contratistas. Half the union members there were employed "de planta;"

---

*An excellent discussion of the nature of landless work opportunities can be found in the essays in *Casual Work and Poverty in Third World Cities*, edited by Bromley and Grey. Although oriented to urban situations, these are highly applicable to employment in regions like Tabasco where, due to good transportation systems and geographic proximity, rural people form a large part of the work force for urban activities.

the rest were transitorios who were hired by the well or for 90 days. The chief of personnel in Cunduacán believed in 1980 that at most ten percent of the petroleum work force was from Tabasco--fewer de planta, perhaps as much as 20 percent of transitorios workers. About a fifth of the people working for PEMEX in 1980 had also been working for the company in 1970. Most of the rest were people whose relatives were former PEMEX workers in Veracruz. There was a big increase in new members from Tabasco in 1979-80.

For other types of jobs, specialized individuals arranged employment. The "enganchador" charged the hiring company payment for each qualified worker he brought in. This system was also used in cane-cutting and some other agricultural activities. The "contratista" arranged with the company to do a particular job, then himself acted as "patron" to the workers he hired, called "destajistas." There was no direct contact between company and workers in this case.

Family members sought work over a wide geographical range, if it could not be found locally. This included travel to far parts of Mexico and the United States, although experience in the United States was more common among older individuals. Migration statistics for the nonfarm families showed very high levels of migration for the age group 16-40, and very low levels for those 51-80, substantiating the reported tendency of the young to migrate (often with their families) in search of better employment opportunities.

Worker Organization

By far the vast majority of landless workers were unorganized and sought work and arranged working conditions individually. There was, however, a surge in unionization and worker organization during the 1970s.

The League of Agrarian Communities, an offshoot of the CNC, organized workers, ejidatarios, and employees to demand better wages and provision of benefits such as social security. Some ranchers and large cacao farmers did participate in these programs, but they were a minority. The focus of the new League program was on urban occupations filled by campesinos. They wanted to strengthen their bargaining position so that transitorios could not be dismissed at their employers' will, and so that contracts would be lengthened to permit payment of legal benefits.

All workers in civil construction belonged to the CTM or CROC unions. Even when hired as libres, they became union members automatically. Only the PEMEX union required workers to be members first. The constructoras did not have to use only union personnel. In Cunduacán in May 1980, of the 5,000 "de planta" union members, 70 percent were from the municipio and the rest from out of state.

Standard of Living of Landless Workers

A landless worker who was fully employed, working 250-280 days per year at $120 per day, earned $30,000-34,000 yearly in 1980, or US$1,350-1,535 (less than half that after devaluation). This income was considerably below that of all cacao farmers surveyed (as seen in Chart 4.1, p. 121). Even adjusting for two full-time workers in the family, income barely reached that of the minifundistas. This resulted in a fairly precarious economic situation because of the rapid inflation in consumer goods prices, especially for food. The value-added tax begun in the 1970s further decreased the buying power of the landless. Many landless families rented small maize plots, but rent for these also went up. The abundant fishing and hunting, an important income supplement not long ago, had disappeared in all but a few areas of the cacao heartland.

There had been offsetting factors, however. Most important was the abundance of nonfarm income supplements available, particularly related to construction. Also, about 20 percent of women in landless households earned some income via formal jobs, trading, or provision of food, sewing, cleaning, or laundry services. One sign of improved welfare was the reduction in female wage labor on the cacao plantations. Most children of the landless in the cacao areas now finished primary school. Though this still entailed great sacrifice to the family, the result was a major shift in employment potential.

The increase in urban occupations, and especially in work for PEMEX constructoras, had led to life-style changes. Many construction workers were gone during the week and returned on weekends. Where travels ranged farther, money was sent to the family at intervals. But there was a distressingly high incidence of family abandonment as well.

The daily schedule of jornaleros, factory workers and Villahermosa construction workers differed considerably. The jornalero woke at 4:30 A.M. like the others, worked from 6:00-9:00 and 9:30-11:00 A.M., lunched at noon and then rested, visited, or did household chores until supper at 7:00 P.M. The factory worker rose at the same time but spent more time traveling. He worked from 7:00-12:00 and 12:30-3:00 P.M., taking lunch at noon and supper after getting home at 4:00 P.M. When working overtime, he often did not get home until 6:00-7:00 in the evening. The Villahermosa worker labored 7:00-9:00, 9:30-12:00, and 12:30-3:00, eating his big meal at 5:00 P.M. after returning home.

The physical demands of PEMEX-related jobs were often greater than those of an average day in field work and, because of longer hours, were more exhausting. Such jobs were also associated with the greater risks described in

Chapter 2, such as explosions and chemical burns. In choosing between such work and agricultural jobs, workers weighed these factors, including lost transportation time, against the higher wages, overtime pay, occasional benefits, and greater status of petroleum work.

CHANGES IN LANDLESS WORKER EMPLOYMENT IN THE 1970s

Four factors combined to change the employment patterns of landless rural workers in the past decade: increasing numbers and proportion of landless workers in the working population; tightening of the land market; increasing prices for local agricultural products; and the great labor demand of the petroleum boom.

Trends in the "Supply" of Landless Labor in Tabasco

The EAP in Tabasco whose principal occupation was in agriculture grew from around 100,000 in 1960 to 116,000 in 1970. The number of farmers alone grew by 10,000, accounting for most of the increase. During this period, most of the remaining frontier areas of the state were settled and the large family cacao holdings were first broken up to accommodate children who had no possibilities of settling on new lands.

Ejidalization did not begin in force in Tabasco until the 1950s, and during the 1960s there were still many ejidal extensions being made. In the 1970s, by contrast, little land was distributed, despite there still being large areas of privately-owned land in the coastal, southern, and eastern parts of the state which could be so utilized. The petroleum boom reduced the urgency of such action, by diverting landless workers to urban-type activities.

The 1970s again saw an increase in the number of farmers, but this was largely a result of minifundization, and thus there was a simultaneous large increase in the number of workers. For the first time in Tabasco, there existed a large group of landless workers who were not part of a landowner's immediate family, waiting to become landowners themselves upon inheritance, or saving money to purchase unimproved land.

Tabasco was making the transition from a state dominated by smallholder farms to one dominated by the landless. In 1960 and 1970, workers and unpaid family workers accounted for half of all agricultural EAP. Whereas farmers accounted for 34 and 30 percent of all state workers in 1960 and 1970, respectively, in 1980 they accounted for closer to a quarter of the working population. The average farm size in the agricultural zone of the state had grown considerably smaller.

These same processes led to tremendous upward pressure on land prices, exacerbated by high agricultural product prices and PEMEX activity, which in turn led to speculation, greater urban demands for land, and general price inflation. A fundamental change in possibilities for the landless had taken place. In 1975 a hectare of unimproved land in the petroleum/cacao areas may have cost $15,000; in 1980 it cost $80,000. Cacao land which formerly cost $25,000 per hectare could cost up to $300,000.

This price increase had effectively destroyed access to land for agricultural jornaleros, who in earlier times could often look forward to purchasing their own land after years of hard work. The earlier land abundance had been one of the principal reasons for minimal class conflicts and fluid social interaction within the cacao sector. Land acquisition was now possible only for individuals who found nonagricultural types of high-paying work, for example in petroleum or skilled trades.

Trends in the "Demand" for Landless Labor in Tabasco

Economic conditions in the 1970s brought some compensation to the landless. The large increase in labor demand induced by the agricultural boom, the state infrastructure development, and the petroleum boom increased labor scarcity. As a result, the proportion of agricultural product price which went to labor appears to have increased.

As discussed in Chapter 3, the minimum wage in Tabasco in agriculture rose faster than in other sectors. Also, the ratio between the agricultural minimum wage and wages for relatively highly skilled jobs (Level VI) narrowed from .43 in 1972-73 to .47 in 1975 and .56 in 1980. But the minimum wage did not accurately reflect conditions in the cacao labor market. The minimum wage was generally paid for a five-hour day, whereas a full day cost closer to $150 in 1980. By contrast, the wage in ranching was much lower, with ranchers paying only $50-60 in 1979 for a five-hour day. The ratio between the price of cacao and the price of a day's labor in the cacao plantations rose from $20/day:$30/kilo in 1970 to $40:$65 in 1975 and $150:$50 in 1980.

The effect of PEMEX activity was important, but should not be overestimated. At its peak, over 50,000 workers were employed on PEMEX projects, but the majority of these were from out of state. After a period of a couple of years, most were gone and with them much of the urban bloat of Cárdenas and Villahermosa. There was not much permanent migration to the state that would compete with local workers. Many of those who did migrate had specialized skills which (had Tabascans been trained for them) would have been superfluous at the end of the "boom." Thus, the influx did not have a depressing effect on local wages because newcomers took newly-created jobs.

The thousands of Tabascans who did participate contributed to the labor scarcity within agriculture and were responsible for significant money transfers to rural areas. Most workers were involved only intermittently. This was in part due to the restrictiveness of the STPRM, the petroleum workers' union, which cost a small fortune to join. Most Tabascan workers were limited to "transitorio" positions with contracts of only weeks or months at a time.

In light of the tightness of the labor market during the boom, the employment objectives of the State Agricultural Plan seem naively optimistic. Short-term objectives included generation of 15,000 permanent new jobs--an increase of over 12 percent in the agricultural labor force, or three percent per year (COPRODET 1979d, p. 116). Most technical assistance programs relied on technology requiring heavy infusions of labor inputs. The Chontalpa Plan doubled the number of hours worked on the collectives during the late 1970s.

Yet agriculture would probably not have been perceived as suffering from unusual labor scarcity at the peak of the petroleum boom had it not been for high commodity prices. These production incentives, and the availability of labor-intensive technologies for intensification, increased farm demand for labor markedly over the level required in the early 1970s.

Tabasco was *not* experiencing a temporary labor scarcity in the midst of a long-term tendency to labor abundance. Data which are cited to support this position (Allub and Michel 1980, Chap. 7) ignore secondary occupations, the role of family labor in production, seasonality, and volatility of employment with agricultural price changes. In fact, Tabasco has traditionally been a labor-short area. The state experienced significant unemployment, mainly in areas newly switched from agriculture to livestock production, and during periods of depressed agricultural prices.

Will a collapse of petroleum development change the countryside from labor-scarce to labor-abundant? Perhaps, but the determining variable will be changing conditions within the agricultural sector itself.

Role of the Landless in Tabasco State Development

The slowdown of the petroleum boom in Tabasco arrived several years before the anticipated date, as a result of curtailment of PEMEX construction activity after the economic "crisis" of 1982. While improved economic circumstances and rising prices for petroleum may lead to another boom in the future, it is likely to be neither so dramatic nor so disruptive as that of the 1970s. The negative impact of the PEMEX slowdown on landless employment opportunities was not as serious as may appear at first glance, because much of the petroleum company labor force was imported. Reduced construction by the State Development Plan in other economic sectors would have had more serious consequences for the landless, but formulas for distribution of revenues from petroleum ensured a generous share for the Tabasco state government, whose investments in other sectors 1982-85 remained very high. Still, the most important factor in the state market for "unskilled" labor continues to be demand for labor from the agricultural sector.

As Tabasco's economy acquires a more diversified structure, serious retraining programs for agricultural laborers must be established--training in the skills required for tourism, for agricultural infrastructure maintenance, for housing construction, for commercial activities, and for the professions. The state university was overextended in 1980 and suffered both from a shortage of teaching personnel and from students ill-prepared in secondary school. Tabasco had succeeded in greatly increasing its number of students; it then needed to attract higher quality teachers who could provide them with acceptable educational standards. The state had several skilled trade training centers in the early 1980s, but they were still largely showpieces and did not produce enough trained people to make much of a dent. Broad provision of such training is essential.

Meanwhile, public policy can influence the terms of access to land for the landless. The potential for land reform in Tabasco is by no means exhausted. Increasingly, large extensions of good agricultural land are used for grazing, which could be rented or distributed to the landless. Since minifundia have a more limited potential for increasing the labor intensity of land than do family-farm-size holdings, policies to counter minifundization should be promoted.

Regulation regarding urban development could be loosened to promote a freer atmosphere for the casual and part-time economic employment which is so critical to the landless. Government contracting could consider the possibility of providing certain types of urban services through small-scale local providers, rather than large-scale national companies which tend to be far more capital-intensive.

EVOLUTION OF THE RURAL LABOR MARKET IN A PETROLEUM ECONOMY

The petroleum boom in Tabasco brought some major, permanent changes to the state labor market: an increased number of nonagricultural jobs, an increase in the general wage level, and an increase in state investment levels. Even with continued high agricultural growth rates, the share of agriculture in state employment is likely to continue to decline. As this study has shown, the apparent 33 percent decline in the official agricultural EAP 1970-80 greatly overestimates the actual exodus from agricultural work. Nonetheless the trend to off-farm employment will certainly continue.

What would the labor market have looked like without a petroleum boom? It is likely that high prices for cacao would under other circumstances have led to much higher total state production of cacao--because of the much increased capital and effort that hired-labor-based producers would likely have put into their cacao farms. Similarly, minifundistas might also have increased their production of cacao substantially, the family-labor-based farms somewhat, and the collectives probably very little.

Still, the production performance of the cacao sector in Tabasco contrasts remarkably with that of a similar cacao-producing region in western Nigeria during the petroleum boom (documented by Sara Berry, 1982). There the surpluses from cacao farming were used almost exclusively to leave farming and find urban, off-farm sources of livelihood. Total production and export levels declined sharply.

Differences between the two regions are difficult to measure because of lack of comparable regional statistics in Nigeria. Smallholder cacao production dominated there, but it is unclear what proportion of producers were dependent upon hired labor; the degree of minifundization is also not clear. The Nigerian cacao farmers were also located nearer to major national urban centers, and there was a stronger on-going pattern of outmigration prior to the 1970s oil boom. In Tabasco, by contrast, cacao farm family workers had a tradition of working off-farm largely in agriculture and in the cacao sector. In Nigeria there were no comparable farmer organizations, no major state programs to assist cacao farmers, no major local alternatives for nonfarm employment, and no major domestic market for cacao. Berry comments that "oil development did not alter the underlying strategies of mobility and accumulation; it just accelerated the rate at which people left the cocoa sector" (Berry 1982, p. 3).

It is possible that Tabasco was simply at a different stage of "proletarianization" and land consolidation than western Nigeria, and that the experience of the latter will be Tabasco's in the next generation. In 1980, the pivotal family-labor-based farms represented just over a quarter of cacao producers. If the dual processes of "minifundización" of family-based farms and "ganaderización" (transformation of cropland to grazing) of hired-labor-based farms continue, then much of the dynamism of the cacao sector may disappear. However, this is by no means certain. Changing distribution patterns of land ownership, transfers, and usufruct rights are not at all well understood or documented in Tabasco, and this is a key area where in-depth field and archival research is needed. In these analyses, the effects of family and political relationships of buyers and sellers, donors and recipients are particularly essential to discerning long-term trends.

Another key area for study is the transition from part-time to permanent rural emigration. In Tabasco a large share of emigration was seasonal or a part of a particular life stage of an individual whose principal long-term source of livelihood is the farm. How stable is this pattern?

Verduzco (1984) outlined four patterns of migration found in Mexico and related them to the type of land and product. He suggested that commercial monoculture, where smallholders are required to plant the crop and subsistence food production perforce declines, led to gradual permanent emigration. So did the transition of land to livestock production, for which labor requirements per unit of land were very low. Areas of poor quality land displayed a pattern of temporary migration, as the land could not produce enough income to maintain the family, but there was no competition for the land from capitalist agriculture. Only on commercial production subject to appropriate innovation were campesino producers retained. This latter pattern perhaps corresponds to the Tabascan situation.

Pucciarelli (1983), in his analysis of agricultural development around the industrial enclave at Las Truchas steel works on the Pacific coast of Michoacán, believed that the trend to high-value tree crops production would displace campesinos with capitalist-style producers. Yet his evidence of investment in machinery to raise soil productivity rather than to save labor, and of increasing campesino participation in the new production regimes, does not convince the reader that this is necessarily imminent.

Far from being another "enclave" like petroleum, Tabasco's export-oriented cacao sector had extremely broad and positive economic integration with the regional economy. Most of the "campesinos" there did very well during the 1970s, including the semiproletariat. Although there is a clear distinction between the campesino whose life cycle stage and patterns of accumulation had allowed him to amass enough cacao land to require hired labor, and the traditional large-scale cacao farmer, the long-term structural results will depend on how land is transferred in the next generation. In the major regions of good cacao land, it is as likely that the large-holders will sell land to smallholders in order to take advantage of other investment opportunities, as it is that further land concentration will take place.

Meanwhile, the cacao-growing families service the economy in two complementary ways. First, they form a huge reserve of potential new workers. Second, they can act as a sponge for absorbing excess workers from the economy.

Such variables have a major impact on the evolution of agricultural labor demand and supply in most petroleum economies, as do the trends in agricultural prices, which determine the degree of intensification and hence labor use. While the petroleum sector influences many of the constraints within which the rural labor market operates, it does not predetermine its structure, nor the resulting agricultural production response.

9

Agricultural Development in a Petroleum Economy: Lessons From Tabasco

By the international standards of other petroleum-exporting nations and regions, the well-being of Tabasco's rural economy was truly notable. One cannot expect that demographic, political, structural, and economic forces will coalesce in such an unambiguously positive way in many other places. But Tabasco's experience in the 1970s does demonstrate that agriculture-petroleum conflict is not inevitable. Whether the disruptive effects of a petroleum boom or similar major industrial transition can be mitigated for the agricultural sector depends on control of macroeconomic "oil syndrome" pressures, appropriate agricultural pricing and sectoral spending policies, supportive regional development policies, and conducive regional attributes of labor demand and availability. This chapter provides some guidelines for developing public policy at each of these levels and discusses the special role that smallholder agriculture is likely to play in successful adaptation of the rural sector to a petroleum economy.

MACROECONOMIC POLICY VARIABLES

Most oil-exporting countries have compelling reasons to maintain a productive and prosperous agricultural sector, to avoid excessively high food imports as domestic demand increases, and to protect the economy from excessive disruption when temporary declines in oil prices abruptly reduce foreign exchange availability. "Food security" (though not food self-sufficiency) is likely to be a guiding policy objective.

Obviously, one of the most effective means of controlling disruption of the agricultural sector is by limiting the flow of foreign exchange and the level of public and private spending through conservative fiscal, monetary, and oil export policies. But so long as "oil syndrome" pressures on relative prices continue, significant intervention in the economy will be necessary to compensate rural producers and workers. Two specific policy areas are particularly important in determining the response of the agricultural sector to petroleum development--agricultural price policy and patterns of sectoral spending.

Agricultural Prices

The greatest single boon to the Tabascan farm sector in the latter 1970s was the highly favorable price situation for major farm products, particularly cacao, cattle, and coconut. The exception proving the rule was the dismal return for maize farming. The maize sector, in fact, conformed to the classically described negative farm response to a petroleum boom.

In order to assure continued agricultural production in a petroleum economy, agricultural income must be roughly competitive with income-earning opportunities outside agriculture. Product prices must cover the costs of at least enough production to support the farm family, even after a decline in (presumably now more expensive) input use. This level will depend on farm budget structures, and sensitivities to items whose price and availability are affected by the "boom." It will probably *not* be the same for all producers of a particular commodity.

High commodity prices encourage production not only for the obvious reasons of increasing returns to family labor and resources, but also through their psychological impact on farmers.* With high prices, farmers are more likely to continue farm investment despite the boom, whereas low prices may hasten the disinvestment process. Agricultural development in areas dominated by low-priced commodities will be extremely difficult, unless there is a reasonable prospect for a change in land use. In Tabasco, the special market conditions for cacao--a "captive" internal market, high international prices, and a cooperative marketing system which enabled farmers to reap much of the gain from high prices--were instrumental in successful development despite oil boom economic conditions.

---

*For a discussion of the multiple reasons to support high food prices in developing countries, see Lipton (1977), pp. 313-15.

Agricultural product prices, and the relative level of prices for different commodities, are determined in part by macroeconomic factors such as foreign exchange policy, trade policy, food and farmer subsidy programs, and policies affecting the composition of demand through changes in income levels and distribution. The domestic agricultural sector is far better off with a properly valued currency, and general economic policies should place greater weight on this outcome. But free trade in food, particularly in the most common circumstance where the currency *is* overvalued, can be extremely detrimental to sectoral development. Import tariffs and export subsidies on agricultural products can serve to significantly offset the macroeconomic pressures of the "oil syndrome" on agriculture.

If resources permit, the parallel subsidization of staple foods for a targeted vulnerable population is certainly justifiable under conditions of rapid, uneven economic growth and high inflation of basic commodities. This policy could offset the tendency for income distribution to worsen in a petroleum boom, and it will increase demand for the consumption items of the poor, increase producer prices, and thus encourage increased production. Such policies will serve to slow the essentially unavoidable (and in the long term desirable) transition in agricultural land use from staple foods to high-value commodities demanded by population groups newly prosperous as a result of the oil boom.

Fiscal Policy and Agricultural Development

As was discussed in Chapter 1, the role of the state in the national economy grows markedly in petroleum-exporting countries whose governments receive most of the taxes and royalties from production. One of the most crucial decisions affecting "oil syndrome" patterns is the sectoral distribution of those revenues. Excess deficit spending and the consequent inflation can be particularly harmful to the agricultural sector. But within a program of responsible fiscal planning, the agricultural sector and agricultural regions should receive a substantial share of total spending and investment. Tabasco's access to federally-distributed petroleum income, as well as its relatively strong state planning apparatus, was a key element in maintaining viable economic development outside the petroleum sector.

Sectoral allocation of government expenditures is an issue of general importance in economic development; the problems are underlined in a petroleum economy because of the sudden and large increase in available revenues whose distribution can have acute and immediate effects on the entire economy. If, prior to the petroleum boom, a strong argument could be made for investing in the agricultural sector; then, during and after the boom, the argument becomes even stronger, not weaker, as many are tempted to believe.

REGIONAL POLICY CONSIDERATIONS

Macroeconomic policy and performance are not the only, and often not even the major, determinants of agricultural development. "Good" national-level policies will not guarantee the desired sectoral response, particularly in the actual zones of petroleum exploitation. Regional and local policy considerations are equally important, with some of the most important factors being political support for farmers, government interventions in the local economy, local inflation control, and local petroleum company policies.

Political Support for Farmers

In the case of Tabasco, strong producer organization for both political and economic goals, a simultaneity of the interests of powerful largeholder cacaoteros with campesinos, the importance of cacao to the Tabascan economy, and national public sympathy for the threatened farmers, all combined to give the cacaoteros unusual political leverage. The cacaoteros, who had long played an important role in Tabascan state government, were instrumental in designing state and federal government programs to stimulate the cacao economy and promote production advances.

Such political strength among smallholder agriculturalists is not common. Even where there is substantial local leverage, it is difficult to build a coalition sufficiently powerful to take on a giant multinational petroleum corporation or state-run monopoly. Where the political will does exist, the regional government can play a crucial role in setting internal terms of trade, limiting petroleum company prerogatives, and providing needed services to the farming communities. Even under circumstances where the structure of agricultural production itself exacerbates the vulnerability of the farm sector, government action can make an enormous difference in terms of the human cost of industrial transition. Well-designed policies can help the farm sector weather an intense, but temporary "boom," enabling agricultural production to adjust successfully when the boom is past.

In Tabasco, the state government sought to control the negative aspects of petroleum development through its inflation control programs, by supporting farmers in expropriation procedures, challenging environmental safety threats, and offsetting farmer insecurity through strong support of agriculture. Benefits from petroleum development were spread chiefly through far-reaching education and health programs, COPLAMAR, PIDER projects, tourism development in outlying areas of the state, and new agricultural infrastructure and development projects.

There may be a tendency to concentrate government infrastructure investments during industrial transition completely in the towns, since this is where shortages appear most obvious. But over the long term, investment in rural development--including schools, housing, sanitary infrastructure--can reduce internal migration and encourage agriculture and can prevent the drastic structural changes in the economy which are the cause of many of the urban problems. Since, under dynamic economic conditions, processes of marginalization tend to speed up, decisions to postpone investments in marginal areas by even a few years (e.g., to postpone construction of farm-to-market roads in favor of urban highways) can lead to severe economic stagnation in those areas bypassed.

Inflation Controls

Accelerated inflation nearly always accompanies a petroleum boom. Additional upward pressures on prices in the producing regions result from sharply increased average incomes and the additional multiplier effect on demand of regional investment projects paid for by oil revenues. Inflation plays havoc with the local economy, particularly those sectors not linked to the leading petroleum economy with its high wages. For farmers, increases in cost of living and production inputs are the leading causes of farm disinvestment and migration to petroleum employment. In Tabasco, the agricultural boom itself--increasing farm incomes, farmworker wages, and total employment--added substantially to local inflationary pressures.

Some of the approaches to reducing inflation in the state were selective retail price controls, government-sponsored competition with private markets, and "bottleneck-breaking" investment. "Competition" was provided through the proliferation of government-sponsored rural stores, opening of medical clinics, assistance to producer cooperatives in acquiring production inputs at cost, and government bus services competing with private companies. Key investments included, for example, a new transport center and new public markets. Policy effectiveness was mixed at best, but in specific zones made a great deal of difference to farm families. Of course, the rise in farm income from rising farm prices was the most important mitigating factor.

Petroleum Company Flexibility

In Tabasco, PEMEX was involved (by federal mandate, and at the insistence of the state government) in state planning board meetings. The company was obliged to inform the state in advance of where and how new developments would be established. While PEMEX participation assisted state planning, the advantages were not all with the state. By sharing information with COPRODET and the Governor, PEMEX occasionally passed to them responsibility for resolving public concerns.

The willingness of the petroleum companies to compromise, consult with, and consider the agricultural sector in their land use planning is of critical importance in resolving agriculture-petroleum conflicts. There is a broad range of issues on which the two sectors are by no means incompatible. Some adjustments would increase costs for the company, but these would be outweighed in real economic terms by reduced losses and added production in the farming sector, not to mention reduced political conflict and litigation. Drainage system protection, pollution control, prior warning of changes, farmland utilization, adequate compensation for expropriated land and damages, mutually compatible road designs and location--all are issues on which petroleum companies can and should reasonably be expected to consider agricultural interests.

Another important factor is petroleum employment policy. Key choices which can significantly affect local labor markets include: transfer of personnel from other petroleum areas versus local hiring; degree of technical skill required for employment versus training availability; use of short-term or long-term labor contracts; educational and other requirements for jobs; unionization, membership rules, and job protection; and determination of pay levels.

In Tabasco, one of the most destabilizing factors for the landless workers and minifundistas was their inability to get contracts lasting longer than three months. (The companies need not pay social security benefits for such short durations of employment.) A lot of time was lost waiting for jobs and little forward planning was possible. On the other hand, the mass transfer of personnel from other states for particular jobs in Tabasco meant that after the boom many migrants left again, reducing the potential problem of mass unemployment.

REGIONAL ATTRIBUTES AFFECTING AGRICULTURAL RESPONSE

Public policies play only a partial role in determining the outcome of petroleum-agriculture interaction. They can be seriously constrained by specific historically and structurally determined characteristics which vary in different agricultural regions. Chief among these are demographic conditions, characteristics of labor demand, labor availability, and structure of the farm sector. While some of these variables are really only of concern in regions close to petroleum activities, most are relevant for all agricultural producing regions in a petroleum economy, regardless of location. Similarly, these variables will partly determine agricultural response to other regional or far-reaching national structural changes, such as industrial development or major technical change in agriculture.

Demographic Conditions

That the principal petroleum zone in Tabasco happened to coincide with the area of greatest population density in the state was a source of serious sectoral conflict. Nevertheless, disruption from rural-urban migration was thereby kept to a minimum. Settlement patterns were such as to encourage continued residence on the farm rather than permanent residence change to the large towns. Population density in the area of petroleum production, regional settlement patterns, and traditional internal migration rates and flow patterns will all affect the nature and severity of agriculture-petroleum conflict in a region.

The rate of labor participation--working age, female participation, the demographic cycle as it relates to working life and local culture--affects the kinds of changes that a petroleum economy introduces. Areas with high traditional female participation on the farm, for example, may prove more adaptive than zones where loss of the male head of household to petroleum employment leaves behind no one trained to run the farm.

Rates and patterns of immigration help determine the acuteness of problems such as inflation, rising wage, rates, and scarcity of housing and other basic goods. Geographical and transport characteristics will profoundly affect government efforts to control the extent to which new petroleum jobs can be restricted to either locals or immigrants. Settlement patterns of immigrants, and their access to social infrastructure, will also affect the degree of social conflict between natives and immigrants.

Labor Demand Characteristics

The flexibility of employment conditions in agriculture has an extremely important effect on the potential for compatibility between the agricultural and petroleum economies. Agricultural labor cycles and seasonality, and the labor sensitivity of different crops, are key variables.

Cacao production in Tabasco utilized labor very abundantly but flexibly. There were few operations for which timeliness was critical, and seasonality of labor use was not marked. Peaks occurred at the height of the rainy season, when construction activities for PEMEX, the government, and nearly everyone else came to a halt. This permitted an influx of multi-sector workers to cacao plantations during the most critical periods.

In general, perennial crops with flexible labor requirements, such as cacao, coconut, and oranges, are the most conflict-free. Perennial crops like avocado and sugar cane, which have definite timing requirements for inputs and extreme labor peaks, are less so. Production of some annual crops may fit easily within the periods of slack petroleum labor demand; others will directly conflict.

A final factor in determining labor demand, frequently overlooked, is the sensitivity of farm employment to agricultural product price. Labor requirements for a particular crop are not fixed over time; when prices are high, more labor will be required for more intensive production, and when prices are low, labor use will generally be reduced. Farms with different management objectives and constraints will respond differently to such price changes, but some change in production and employment patterns is certain.

Labor Availability

The structure of the EAP, the relative importance of agriculture as an employer and producer, levels of regional unemployment, a custom of off-farm employment, and the sensitivity of employment to agricultural product price are other important variables affecting petroleum-agriculture interaction.

Labor availability for agricultural production is dependent on a number of factors. The EAP occupied in farming and farm work is itself dependent on cultural characteristics such as the role of women and children, structural characteristics such as land tenure patterns, and demographic characteristics such as the age distribution.

Whether the EAP on farms and in rural areas will choose to work on farms depends on what their alternatives are and on cultural work preferences. In a petroleum economy, the alternatives may be both financially and personally more rewarding. Regional unemployment levels, and existing patterns of off-farm employment, will also affect these decisions. The more flexible the time requirements of off-farm jobs, and the better they coordinate with agricultural labor demands, the more likely are farmworkers, and particularly farmers, to seek them.

Structure of the Farm Sector

Most of the demographic and labor market characteristics of a regional agricultural economy stem fundamentally from the "structure" of the farm sector--the distribution of ownership and usufruct rights to land; the agronomic constraints on the feasibility of different types of farming systems; and the productive and social relationships between different types of farm households, different members within farm households, and different farming systems. To determine regional policy needs and effects, it is necessary to trace patterns and flows of land rights, capital, labor, and other key agricultural inputs through the existing farm structure.

AGRICULTURE IN A PETROLEUM ECONOMY

Petroleum development does lead directly to changes in the agricultural economy. This section will summarize some of the key expected land use changes, the potential role of smallholder producers in a petroleum economy, and agricultural development policies appropriate for the new economic conditions.

Structural Change

Petroleum development induces structural change in the agricultural sector through major shifts in resource availability and relative prices, and shifting demand and production patterns. Indeed, the enclave nature of petroleum production that apparently used to characterize oil-based development has changed substantially now that a far higher portion of petroleum income is retained by producing countries and regions. Even in cases where the technical demands of petroleum production do not themselves have large domestic forward and backward linkages, the multipliers from petroleum worker income, related construction, and government spending of revenues can be huge. The extent to which agricultural producers will either participate directly in such activities, or revise their production patterns to meet petroleum-economy derived demand or pricing structure, will depend on the pattern of distribution of oil revenues and the structure of agricultural production.

Where there is high demand for workers without sophisticated skills, the wage level outside agriculture will rise, potentially attracting both farmers and landless workers. The extent of exodus from agriculture will depend upon the differential between agricultural incomes (which are not easily calculated) and nonagricultural incomes. It is important to recognize that for smallholders the decision is not an "either-or" choice. Rather, decisions address the relative importance of various activities (farm and nonfarm) through which family cash needs will be met. The composition of farm production will depend upon relative prices and costs of production. As national incomes rise, there will certainly be a huge increase in demand for food, especially of non-staples such as fruits and vegetables, meat, and dairy products. A shift toward production of these commodities should be expected and is often desirable.

The key information needed to predict or analyze the nature of such shifts includes:

- changes in *net* farm income derived from different products (a function of changes in relative farm prices and changes in relative farm input costs)
- availability of technology for intensification of production, and the cost of farm management changes needed for yield increases for different products
- the agricultural structure, and changing conditions of production for different types of farmers in different places (i.e., the kinds of farmers who are in a position to make these changes)

During the 1970s in Mexico, the rate of growth in high-value crops was very high; that of maize and other staples was relatively low. These who did *not* move out of maize production during the periods of very low relative prices were those on land too marginal, or with capital resources too limited, to produce the higher-value crops. "Campesinos" who were on more fertile lands *also* switched to higher value crops.

Indeed, much of the debate in Mexico over high-value crop production and its role in the demise of the campesinado, is due to preoccupation with land use and products, rather than producers. For example, substitution of subsistence and staple crops production by higher-value crops production took place throughout the tropical regions of the country that could not produce maize for sale as cheaply as other parts of the country. But in most cases, this reflected neither the *replacement* of campesinos by capitalist producers, nor their *transformation*, but rather new cash strategies through commercial production of crops providing higher returns to family labor. In areas already dominated by largeholders and capitalist producers, there was frequently a tendency for *reduced* intensity of production and accelerated reversion of land to lower-cost grazing.

There are certainly large areas in other parts of Mexico where land concentration by capitalist producers has led to exodus by smallholders who did not have access to the necessary resources for higher-production cost crops. But this is a result not of the product shift per se, but of the gross maldistribution of resources in the countryside. SAM was a program conceived initially (and was most appropriate) for agriculturally marginal regions supporting large campesino populations whose production was oriented largely to subsistence needs. A similar program needs to be developed for smallholdings in more fertile regions, providing capital and organization for smallholder cooperatives for higher-value products. There is no intrinsic reason (apart from the temptation of governments to encourage it) that agricultural structure in petroleum economies should shift away from smallholder agriculture, despite shifts in product composition.

Re-Examining the Role of the "Peasantry"

The Tabasco case study indeed illustrates the strong case for smallholder-based agricultural development, contrary to commonly-expressed concerns of both neo-Classical and neo-Marxist economists.

The process of economic growth has often brought increased concentration of landholdings, displacement of many campesino producers by capitalist producers using capital-intensive technology, and proletarianization of agricultural producers. This is likely to continue. But in Mexico, as well as in most other developing countries, the smallholder mixed subsistence/commercial farming sector is *not* likely to "disappear" in the foreseeable future. The campesinado provides a reserve labor force for other sectors and raises and supports individuals in a low-cost (relative to the broader economy) environment. Their support system acts for the family as nursery, sickroom, and retirement home--relieving employers outside the campesino economy from having to meet these costs in the form of higher wages or other protections.

The preservation of a sphere of family activity directed to subsistence is a stable form of organization in societies where instability and uncertainty are great, market circuits operate poorly, and industrial development generates little employment. Farmers' insistence on a subsistence sphere of household production, far from being backward, may be the most rational organization possible--a positive adaptation to economic conditions (Greenwood 1978; Merrill 1984).

As nonfarm employment opportunities grow, the degree of smallholder participation in such activities will fluctuate according to relative returns to farming and off-farm jobs. Such participation in off-farm labor markets is the result of changing "cash strategies"--not necessarily a sign of the demise of such systems of production. The need for these farmers to pursue a cash strategy as well as a subsistence strategy means that agricultural production programs for cash crops, if organized in such a way as to be superior to other alternatives for farmers to acquire cash (such as off-farm labor), and properly cognizant of subsistence concerns, will meet with enthusiasm from these producers.

My bias in favor of smallholder production structures in a petroleum economy, and other highly unstable economies, is partly predicated on the high and inflexible cost structures inherent to larger-scale enterprises. These characterize both large private holdings that are in direct competition for capital with nonfarm enterprises, *and* collective agricultural holdings that face different--but no less constraining --problems with capital and labor management.

While smallholder systems are by no means homogeneous, most have numerous advantages over largeholders (as agricultural producers) in an unstable economy: lower fixed costs; flexible labor supply from family members; more flexible timing schedules for agricultural activities (due to smaller scale); less mobile capital; capacity to operate by "self-exploitation" (with lower labor costs per worker than the comparable wage rate); limited role of hired labor in production; willingness to engage in substantial labor mobility on a temporary basis; and exclusion of capital value of land as a production cost (except in the case of rented land). One can find numerous examples of cases where smallholder production systems have supplanted formerly dominant large-scale capitalist producers during economically unstable periods (Merrill 1984, pp. 485-86).

Furthermore, the experience of developing oil-exporting countries suggests that large-scale, capital-intensive, enclave agricultural development in such economies has been terribly disappointing and costly, while the few cases of development policy geared more toward smallholders--Indonesia, especially, and Mexico itself--must be seen as successful by comparison (Scherr 1985). The preponderance of evidence suggests that smallholders demonstrate superior land productivity and efficiency (Berry 1984; Shiras 1978; Johnston and Kilby 1975). Where agricultural land resources are limited and agricultural labor abundant, a smallholder-based agricultural structure would seem preferable. Only under conditions where shadow-priced technology is shown to be both far more productive with large-scale production and unsuitable to cooperative or otherwise organized access to smallholders, ought a largeholder-oriented program be considered.

Agricultural Policies

The terms of smallholder access to key factors of production can be altered substantially through specific agricultural development policies. Public investment programs need to emphasize small-scale agriculture directly, particularly by providing social and agricultural infrastructure. Programs that disrupt on-going economic patterns or make abrupt changes in production systems should be avoided. The principal concern of policymakers should be to maximize the stability and security of the farming sector, not to introduce still more uncertainty. Development programs need to reinforce farmer flexibility, not restrict their options.

Institution building should focus less on independent project-oriented organizations and more on those which provide services to smallholders to compensate for the specific disadvantages of small producer scale (bulk purchase, technical assistance, marketing, for example). Subsidization of one or two key inputs for which supplies are adequate, and that have been shown to markedly increase production on small farms, should certainly be considered. But blanket subsidies can be very expensive, difficult, and disruptive to reverse during periods of low government revenues. Marketing efficiency of inputs and outputs should be emphasized, which may sometimes mean leaving certain markets entirely or partially in the hands of private enterprises or local cooperatives. In matters of input distribution (be it fertilizer, credit, or technical assistance), the goal is broad availability. As has often been asserted, subsidizing inputs that are already scarce will further increase their scarcity for smallholders.

Offsetting oil syndrome bias against agriculture is particularly difficult under conditions when, in addition to all the other problems, productivity outside agriculture is increasing much faster than in agriculture. Agronomic and farm management research that develops technology to increase the productivity of agricultural lands will favor smallholders. "Packages" must be designed that require more limited amounts of credit and cash resources and that build on existing smallholder management systems.

Before technical recommendations can be made, there must be at least some minimal analysis of whole farm systems; single crop budget analyses can be very misleading. This procedure need not entail complex computation or methodology. Simple small-scale surveys can be utilized that highlight potential conflicts between input requirements for different products and that reveal general farm expenditure patterns and the relative importance to farm costs and family income of the crop in question. Key factors would be the role of family versus hired labor in agricultural activities and the importance of overall labor costs in production. Such analysis would, as a matter of course, incorporate information on major nonagricultural income sources, expenditures, and employment priorities. Since the farming population can clearly not be assumed to have homogeneous production characteristics, policy recommendations must be oriented to specific groups of farmers, based on budget analyses of their economic conditions, capacities, and special advantages.

USING RURAL ECONOMY ANALYSIS IN DEVELOPMENT THEORY AND POLICY

One final recommendation which stems from the Tabasco study addresses a major intellectual problem confronting those who study or practice development economics, particularly in agriculture. Peasant economies and their role in development are at the heart of some of the great theoretical debates--both political and economic--of our age. The unfolding of these debates, however, has proceeded at a rarefied level. While they help to sensitize the researcher or policymaker to certain problems, the theories often provide little guidance for the design of useful field research or for the collection of key information with which to make practical policy. Indeed, field research carried out to explore macrotheoretical issues has more often been designed to *validate* rather than test or elucidate those theories.

In the study of Tabasco this was a particularly serious problem for me as a researcher. There were two main bodies of theory relevant to the Tabasco problem of agriculture-petroleum conflict. The first was the "oil syndrome" literature derived from the neoclassical school that analyzes the change in relative prices between agriculture and the petroleum economy, and predicts declining relative returns to agriculture, reduced production, and accelerated rural emigration. The other was the neo-Marxist analysis of the impact of capitalization on the peasantry that predicts accelerated proletarianization and the demise or relative decline in importance of smallholder farming.

That neither of these scenarios occurred in Tabasco during the 1970s--indeed, their converse was the case--does not disprove either paradigm. But neither literature provided a guide to the types of information that needed to be gathered in the field in order to test, qualify, or modify the theoretical analysis under specific economic conditions. Field work provides an essential opportunity for researchers to test hypotheses, but this is only possible if there is a systematic attempt to compare a broad range of alternative explanations of reality on a fairly sophisticated level. We need better methodologies for setting out these alternatives.

Furthermore, if field research is to make significant contributions to theoretical work (which I believe to be essential), then we also need to improve our methodologies for systematically linking field studies to aggregate economic analysis. As outlined in Appendix A, this study made a modest attempt to develop such a research method. Though the purpose of the research was to answer broad questions of agricultural development in a petroleum boom, the study was grounded in the detail of development. The specifics of history, ecology, economy, and demography were critically important for the interpretation of broader issues. To predict or analyze the path of agricultural change in other petroleum economies requires at least cursory review of these variables. The outcome will vary: it is this variation--between countries, regions, farmer types --which needs to be described, analyzed, and emphasized. Otherwise, a macro-theoretical perspective can result in excessive reductionism and in theories of little practical use to the policymaker or analyst dealing with specific problems.

CONCLUSION

In Tabasco, powerful groups worked on behalf of the cacao producers. Will state or national governments elsewhere take upon themselves this burden when agricultural producers have little political power in their own right?

That they do will be critical in those areas of the world newly involved in petroleum development, such as the Mexican Southeast and West Africa, which are populated regions of settled agriculture dominated by smallholders. Despite the visibility of the petroleum sector in the regional economy, agriculture and nonpetroleum economic activities will continue over the long term to provide the principal income and employment base, and to determine family income levels and distribution, for the majority of people in these areas. Not the economic characteristics intrinsic to petroleum development, but the quality of agricultural policy and programs, will determine the viability of their agricultural economies.

The potential resilience of the smallholder or peasant sector can be a critical factor in stabilizing petroleum-exporting economies. Will these agrarian nations make the *simultaneous* commitment to smallholder farming and large-scale industrial development necessary to reduce the inherent conflicts? This will be their central development challenge during the coming decades. The history of the Tabascan cacao farmers in the 1970s and early 1980s offers some hope of success.

Appendix A

Methodological Considerations

> *Among the numerous reasons which account for [our not using inductive field methods in socioeconomic research] is the failure to realize that we are so ignorant of the economic condition of man in rural tropical regions that we do not know how ignorant we are.*
>
> --Hill 1982, p. 5

THE RESEARCH PROBLEM

The "modified oil syndrome" analysis suggests that there are four key variables of the petroleum economy which tend to undermine the agricultural sector: an overvalued exchange rate; changing relative prices of agricultural and nontradeable products; burgeoning off-farm employment and rising nonfarm wages relative to farm wages; and urban-biased distribution of large government surpluses. The expected responses from the agricultural sector to these changes include: a transfer of farmers' productive resources away from agriculture and toward nontradeable goods sectors; declining on-farm labor by farmers and farmworkers, and increasing off-farm employment; and declining farm production as a result of the first two responses, with a particularly marked decline in production of export commodities.

The principal research problem for this study was to explain the *mechanisms* of aggregate agricultural sector response to a petroleum boom; explain possible modifications of the expected response patterns; and determine how the pattern of agricultural response in turn affects key macroeconomic variables, particularly employment. I was concerned with isolating the key analytical variables which need to be included in studies of agricultural development in petroleum economies.

For these purposes, Tabasco was an extremely instructive case study, though neither the state economy itself, nor its response to the petroleum boom, was representative of Mexico as a whole. The "boom" factors were all unusually strong there. Its export-oriented agricultural economy was highly sensitive to exchange rate valuation. Nonfarm economic growth was very high, as were nonfarm prices. There was especially high demand for (and scarcity of) nontradeable goods for construction and transport. Unrestricted borders with the rest of the country, combined with higher than average rates of inflation, led to very high commodity "imports." There were marked increases in off-farm employment opportunities in PEMEX, construction, and service activities. There was a very high level of government spending, mainly in cities and mainly outside the agricultural sector. Yet the agricultural sector was extremely prosperous. An explanation of this contradiction was likely to enhance our understanding of agricultural change and policy options in other petroleum economies.

While oil syndrome problems have been defined mainly in macroeconomic terms, a regional study seemed more appropriate for attempting to understand and document farmer response patterns in a systematic way. Tabasco state was indeed somewhat self-contained: trade flows, public policy, and farm producer organization were all organized in a state framework, though its economy was strongly integrated with Chiapas and Veracruz state labor and commodity markets. While the state economy did not experience "macroeconomic" problems in the same way as would a nation, microeconomic responses to economic phenomena were likely to be similar. Farm response patterns to changing prices, wages, and costs could be analyzed for their impact on aggregate state production and employment levels. A national-level study would have to utilize similar information taken from a number of different regions.

USE OF FIELD STUDIES IN SECTORAL AND REGIONAL ANALYSIS

The presumption of this study is that aggregate analysis is not an adequate substitute for disaggregated, field-based analysis. The following section will review the problems of aggregate analysis and the importance of field studies in this kind of research; address issues of how to do limited local-level analysis relevant and useful to aggregate policy issues; then discuss the collection of microlevel data, using a household economy framework.

Problems of Aggregate Analysis

One of the major reasons for uncertainty about the impact of petroleum development on agriculture is that the intellectual and policy debate has taken place almost exclusively around the evidence of aggregate production and employment statistics.

There are two fundamental theoretical problems with using aggregate studies for analyzing agricultural response in a petroleum economy. First, the economic characteristics of a petroleum boom constitute a major structural change; hence prior trends in the behavior of aggregate variables must be suspect as a guide to future behavior. One of the key purposes of field research during such a period is to gauge likely deviations from trends resulting from structural changes.

Second, aggregation assumes functional homogeneity of its components. For purposes of analyzing unknowns--farm response to a petroleum boom--this presumption of homogeneity is not only inappropriate, but possibly highly misleading. Data representing aggregate behavior can disguise marked contrasts in subsectoral behavior, which may have important theoretical and policy implications.

There are also some practical problems with aggregate analysis of farm sector response in developing countries. The extremely poor state of agricultural statistics in most low- and middle-income oil-exporting countries creates much confusion about what is happening even at the aggregate level. Mexico's data, though better than those from many other countries, are still poor.

Almost all economists who have actually worked with large-scale, government-collected sources of data on rural economies recognize their unreliability and incompleteness. Employment statistics fail to capture the complicated labor use systems in agricultural regions; active migration patterns confuse information on population; rural income data is so difficult to collect that one can never even consider using statistics which are derived from only a few questions on a brief questionnaire. Agricultural production data is notoriously difficult to collect even in advanced agricultural systems with well-defined plots and monoculture; in tropical systems of multiple-cropping, with complicated ownership patterns, ill-defined boundaries, continuous harvesting, and underground storage of key products, it is especially challenging. For practical reasons, rules of thumb have been developed for measuring these different variables. But economists often forget that they are only rules of thumb, and assume that we have a much better picture of what is going on than in fact we do.

This process of simplification and denial of severe field data collection problems is almost essential for carrying out macroeconomic and regional economic analysis as it is currently done. We utilize ever more sophisticated techniques for data analysis, which in fact have fairly strict theoretical data requirements. Neither our theoretical nor our statistical tools are well-suited to the mess of agricultural statistics that is the reality in developing countries.

Agricultural production response analysis, which is essential to study of the "oil syndrome" affecting agriculture in developing countries, is a particularly good example. Our most important tools include econometric analysis of price/production and other relationships, and development of regional agricultural "production functions" for specific crops. However, both of these approaches have severe methodological problems for practical application in developing agricultural economies.

Econometric analysis using time series data faces the problem of cycles of land use (planting, etc., for tree crops; fallowing practices); the extreme yearly variability of production of rainfed crops due to weather; and the frequency of major structural changes in basic production factors. Helleiner in the mid-1970s reviewed the production supply response literature from Africa and found it to have been almost wholly unsatisfactory. He noted:

> *to understand the observed correlation--and thus to derive generalizable results therefrom--one clearly requires far more detailed information than one customarily has: data concerning welfare functions that incorporate attitudes to risk, the constraints on existing farming systems, and the interrelationships between various inputs and outputs of such systems in the appropriate environmental context [Helleiner 1975, p. 33].*

Aggregate production functions are not often derived from in-depth field data collection. This approach is confronted with a degree of heterogeneity among producers that makes the choice of a particular farming system for modelling extremely difficult.* Even the less pronounced aggregation often found

---

*Not having personally done research in Asian rice systems, I will take the word of many of those who have, that the degree of heterogeneity is substantially less than that in tropical America or Africa, or in "frontier" regions elsewhere. It has been suggested by several colleagues that the predominance of early economic work in Asian rice-growing communities may explain the general professional acceptance of homogeneity assumptions.

in large scale, detailed agricultural surveys can mask policy-significant behavioral characteristics for different types of farmers.

Sophisticated models of farm household behavior, for example using linear programming, do attempt to measure and explain heterogeneous systems. But because so little micro-level research is available for different farming systems and household economies, such efforts face extremely severe problems of inadequate data. Benito, attempting to model technology adoption behavior in Puebla, Mexico, recommended more research emphasis on the study of *variances* rather than *means*; utilization of socioanthropological case studies; and much more extensive data on farm budgeting and the allocation of human time (Benito 1983, pp. 445-49). Even multivariate models can be misleading if, in fact, there are discrete groups of producers who have very different systems of production, and who are in dynamic interaction with one another.

Our analytical tools have been designed for use with reliable and well-understood data bases. Where we are dealing with production systems which are not well known and have not been quantified, we need to begin any study of farm sector response with systematic field research.

"The Sum of the Parts": Meaningful Disaggregation

Without understanding farm-level behavior, it is extremely difficult to predict farm responses to petroleum economy changes, or to agricultural policies designed to counter oil syndrome distortions. But macro- and sectoral economists have been impatient--and often rightfully so--with a "village economy" approach to regional and national economic problems. While aggregate statistics may be poor in quality, the representativeness of the higher-quality field data from community studies may also be extremely questionable. This has been true of village studies carried out by economists, as well as those by anthropologists.

However, recognition of the problems of aggregation--even in places like the United States where official agricultural data collection is among the best in the world--is now so widespread that attempts are being made to find techniques of disaggregation which will allow the systematic use of more reliable field studies. Just and Zilberman (1984) attempted to classify farm families into "behavioral regimes" to study the distributional impact of agricultural policy in the United States, and to determine differential supply elasticities for farms in different "regimes." Ehrensaft, LaRamée, Bollman, and Buttel (1984) examined microdynamics of Canadian and northern U.S. grain farmers, by following *individuals* through the historical record. They found that small recorded net changes in the number of farms since the 1950s in fact masked extremely large gross levels of farm exit and entry. They attempted to group farmers by different entry/exit behavior and to address the policy implications of their findings. Gladwin and Zabala (1984) developed a typology of farmers based on characteristics determining farmer response to the early 1980s agricultural crisis in Florida. Barlett (1984) found very different levels of economic performance among different groups of farmers in Georgia, with each group needing and responding to different policy initiatives.

A range of potentially useful typologies of farm producers can be derived from what is already known and hypothesized of farm household behavior in the existing literature. Except in farming systems which are very well understood, field research will be required to determine which of the possible typologies will be most useful and any modifications that need to be made in them (see Shaner, Philipp, Schmehl 1982 for more in-depth discussion of relevant farming system research methods).

It is quite clear, however, that the specific typology chosen will only be appropriate for specific research questions. To study the impact of credit changes on supply response, it may be necessary to group separately farmers who have title to their land from those who do not. For a study of the impact of currency devaluation on agricultural incomes, by contrast, it would be more useful to differentiate all producers of export crops from all producers of domestically consumed products. One wants to limit the number of sub-groups considered to those whose differences in behavioral characteristics could have a clear impact on the particular variables being analyzed. One must be cautious about using many of the standard typologies, for example, capitalist/campesino; ejido/propietario; subsistence producer/commercial producer. For many analytical problems, these typologies may be very misleading and obscure more important intra-group differences.*

To be useful in aggregate economic analysis, a typology must also be defined so that it can be linked to existing aggregate economic indicators. "Attitudes," for example, may differentiate producer behavior accurately, but field studies based on such factors cannot be statistically related to the general population. By using a typology based instead on the kinds of data collected in large-scale surveys or

---

*For example, Martínez and Rendón (1978) carried out research which suggested that there were a dozen types of distinct farm household behavior, with "campesino" producers of one type responding very differently to wage and price changes than other types.

censuses, conclusions from the field studies can be extrapolated via census cross-tabulations to suggest the numbers and distribution of individuals in the population at large which can be expected to display specific behavioral characteristics.

Not only must the concepts of producer differentiation be clearly capable of extrapolation, but the choice of population for field studies must be a careful one, so that its degree of representativeness is clear. This is not always easily done. In Tabasco there were literally no published data on numbers of farmers producing different products or occupying different types of land. Census and SARH statistics dealt almost exclusively with agricultural land use, rather than farm holdings. Given the often serious political repercussions that might be expected from more accessible statistics on land use and distribution, many governments will not provide, as a matter of course, such analyses to the researcher. Nevertheless, there are ways of making intelligent estimates. In Tabasco, producer organization membership information, land area distribution data, and production and yield data permitted me to make rough estimates of the numbers of producers in different producing regimes throughout the state. The Sociodemographic Survey of Tabasco (1980) used a simplified but similar approach in choosing rural sampling regions. I was also able to compare my estimates of farmer numbers with those from the above survey, several commodity producer censuses, and agroindustry estimates.

The associated research methodology may be conceived as a "pyramidal" structure.* It begins with a political economy study of the sector under consideration, including structural analysis of the linkages between the regional system and larger socioeconomic formations (for example, the characteristics of export trade), and of the regional socioeconomic system. The latter can be carried out through a systematic "reconnaissance" approach, which attempts to identify significant variability in the region regarding the problem to be studied, and through historical documentation and analysis which shed light on the dynamic qualities of the production systems.

The next level of analysis is of the particular socioeconomic factors which affect farmer response at the local level: market structure; channels which regulate producer access to factors of production and resources such as land, water, or labor; and government development efforts, includ-

---

*The subsequent discussion in this section is based heavily on Merrill (1984, Chap. 2), Greenwood (1978), and Scherr (1984a).

ing education, infrastructure, technical assistance, and credit. Much of this information can only be captured through study of communities. Thus, a key element of the "pyramidal" methodology is the choice of a small number of communities which might be expected for historical, geographical, or political reasons to differ substantially in these characteristics. Not all need be studied in depth, but key findings in one or two need to be at least cross-checked in the others.

The final step up the pyramid is analysis of producer household economies. This level "reveals the mechanics of adaptation as well as the variability and flexibility of responses within the peasant sector resulting from differences in households' access to resources and the factors of production, and from variations in their participation in the capitalist economy" (Merrill, p. 41).

When household economy analysis is complete, one must work one's way back down the pyramid. Household and community-level analyses should enable more accurate interpretation of regional economic interactions and suggest how "the whole" may be more than the "sum of its parts," i.e., how economic behavior of definable groups of producers may interact to affect major economic variables at the regional or national level--prices, production, input demand, employment, migration.

APPROACHES TO FAMILY ECONOMY ANALYSIS

Neoclassical microeconomics has been applied extensively in the past thirty years to studies of production, management, and policy responses of agricultural enterprises in the United States and elsewhere in "developed" countries. Agricultural production economics emphasizes the relation between profit maximization and efficient resource use. Given the high cost of labor in these economies, ample land resources, and the direct ownership and control of large agricultural properties by either individual families or corporations, efficiency generally came to be associated with capital-intensive technologies. The goal of maximizing profits--i.e., the return to capital--became widely accepted.

Farm enterprise studies in developing countries have generally built upon this body of knowledge. They have worked reasonably well for that sector of third world farm economies which approximated those that appeared in the post-World War II, capitalist west--individual or corporate proprietorship on relatively large holdings, oriented toward market production, where relative returns to capital were the guide for farmer investments.

But the methodology breaks down considerably for understanding or predicting behavior of subsistence and mixed subsistence/commercial farmers,* due to their fundamental distinction between "use value" and "exchange value." "Use value" refers to an object's direct utility for the family (apples for eating, furniture for home comfort); "exchange value" refers to a commodity value when exchanged in the marketplace (home-produced apples exchanged for purchased furniture).

In the mixed subsistence/commercial household, some goods are produced for use value (the subsistence sphere), and others are for market (commercial sphere). Labor time and other inputs are not valued the same in the two spheres. Producers may respond to relative price changes in the same way as a capitalist when they produce for the market; but not in the subsistence sphere where meeting minimum subsistence requirements is the guiding rule. Subsistence objectives "distort" resource patterns, and inputs and outputs in the subsistence sphere cannot be priced, since they do not enter the market. Table A.1 outlines the key distinctions between "campesino" or "peasant" production and "entrepreneurial capitalist" production, as set out in a major study of the Economic Commission for Latin America (CEPAL 1982).

With the growing importance of farm management analysis and economic prediction for agricultural prices and production in development policy, new approaches have been devised to deal with these problems. The earliest approach was to develop shadow pricing for the subsistence sphere. These were originally market prices for the nearest input and output substitutes (e.g., pricing labor at the going agricultural wage rate). Although for many purposes this was adequate, the result was often to support the theories prevalent in recent decades that such farmers, judged by neoclassical criteria, were behaving "irrationally" or inefficiently.

Another solution deriving from the microeconomic tradition has been the "new household economics" (Schultz 1974), commonly utilized in studies of industrial economies. It focuses specifically on decision making within the household concerning consumption and production. Its most celebrated

---

*I have tried in this book to minimize use of the term "peasant," or even the somewhat less emotion-laden word "campesino," to refer to these farmers. These terms carry a lot of "baggage," and often imply behaviors, attitudes, and societal roles which are not always accurate. They also imply a homogeneity with which I strongly take issue. I believe that the concept of the smallholder, producing both for market and for household subsistence, is more useful.

Table A.1 Principal Characteristics of Peasant and Entrepreneurial Agriculture

	Peasant Agriculture	Entrepreneurial Agriculture
Object of production	Ensure reproduction of family and the unit of production	Maximize profit and the accumulation of capital
Origin of work force	Fundamentally the family and, on occasion, reciprocal interchange with other units; exceptionally, wage labor in marginal quantities	Wage labor
Working bond of farmer with farm workers	Absolute (family ties)	Nonexistent, except by legal composition
Technology	High intensity of labor; low capital intensity and inputs purchased per day of work	Greater density of active capital and greater proportion of purchased inputs in the value of the final product
Destiny of product and origin of inputs	Partially commercial	Commercial
Criteria for intensification of work	Maximum total product, even at the cost of declining average product; the limit is at zero marginal productivity	Marginal productivity equal to wage
Risk and uncertainty	Nonprobabilistic avoidance of risk; "algorithm of survival"	Probabilistic internalization, seeking profits proportional to risk
Character of work force	Marginal or nontransferrable valorization of work force	Only employs work force transferrable accroding to qualification
Components of income or net product	Indivisible product or family income, realized partially in kind	Salary, rent, and profits exclusively monetary

Source: Adapted from CEPAL, *Economía Campesina y Agricultura Empresarial*, p. 31.

application, which has also been used in the developing economies, is in the decision to have children and the participation of women in the labor force. The modeling criteria in the "new household economics" has been extensively applied to farm decision making in LDCs. Though intrinsically difficult to weight, multiple objective functions are used to incorporate subsistence/commercial production decisions into the analysis. This model takes into account both the variable relation to the market of the mixed subsistence/commercial producers and the centrality of the household, rather than the farm, as the unit of analysis.

The growing concern with issues of subsistence assurance and risk avoidance spawned another new body of theoretical literature on "uncertainty," which also tried to modify the basic microeconomic paradigm to incorporate risk and uncertainty (Berry 1984). New empirical research that overwhelmingly confirmed the basic "rationality" of peasant decision-making behavior led to development of the "farming systems" research approach, which was concerned with the effects of specific environmental, technical, and institutional conditions on problems of rural resource allocation (Shaner, Philipp, and Schmehl 1982). Helleiner had commented back in the early 1970s: "The systems approach to the understanding of agriculture suggests that the ceteris paribus assumptions of orthodox microanalysis may . . . not carry one very far" (Helleiner 1975, p. 43). He encouraged researchers in Africa to focus on farm management studies which shed light upon smallholder decision making and concluded: "It is the *variability* in decision making in curcumstances of change that is likely to shed the greatest light upon the sources of agricultural development" (Helleiner, p. 52; my italics).

Following more than a decade of somewhat inconclusive debate about the appropriateness of using microeconomic criteria of rationality for nonwestern farm producers (LeClair 1968), economic anthropologists began to explore yet another route--the concept of a distinct production logic for peasant producers.

Seminal to this work was A. V. Chayanov's study of Russian peasants in the 1800s. Chayanov insisted that peasant economic behavior was fundamentally different from the capitalist and socialist models. The peasant family was both the unit of consumption and the unit of production; decision making placed priority on subsistence security. The demographic stage of the family--and the ratio between workers and dependents--was responsible for differing levels of production effort, resource use, and family prosperity (Chayanov 1966; Shanin 1971).

Chayanov's specific model of the autarkic peasant family --no off-farm employment, no hired labor, no marketed goods--

is no longer applicable to the vast majority of those loosely termed the world's "peasantry." However, his approach spawned numerous innovators who have adjusted the theory to fit modern conditions (de Janvry 1978; Esteva 1978; Ewell 1984; Greenwood 1973; Hunt 1979; Martínez and Rendón 1978; Merrill 1984; Okurume 1973; Ortiz 1973; Popkin 1979; Scott 1976; Warman 1980; Shanin 1971). My own use of the concept of smallholder mixed subsistence/commercial farming as a particular economic adaptation, with their own objectives and strategies, borrows from this approach.

As empirical studies of subsistence and subsistence/commercial farming systems have grown in number, field anthropologists particularly, but also field economists, have grown increasingly frustrated with the limitations of the microeconomic model in explaining the basic problems of smallholder agricultural development and in developing effective policy prescriptions. The result has been an increase in the use of anthropological techniques to analyze the context within which decisions are made and the importance of "noneconomic" motivations in economic behavior.* Of particular concern are the questions of access to resources and productive inputs and the demands and constraints of social relationships.

Recent reviews of the state of household decision-making analysis outline recommendations for a more comprehensive approach to family economy analysis. Schmink (1984) emphasized the need to include analysis of intrahousehold decision making and to incorporate the socioeconomic relationships of the family with nonresident members. Like Greenwood (1978) and Helleiner (1975), she encouraged a focus on "patterns of variability," rather than "universal" theories.

Berry (1984) criticized decision-making analysis, as commonly practiced, for focusing too narrowly on market costs and returns to alternative crops and forms of employment. She emphasized that while research on rural households is crucial to understanding macro processes such as agricultural performance and employment, in order to do so it *must* "encompass the full range of economic, social, and political relationships among households, and between villagers and extra-village agencies" (Berry, p. 15). Family decision making alone does not explain *how* family options are determined and how they change over time. Issues such as social conflict and relations of power, and strategies to maintain

*An excellent collection of a variety of views and new methodologies utilized by economic anthropologists to study farmer decision making may be found in Barlett (1980). This is essential reading for agricultural economists who utilize decision-making approaches in their research.

and increase access to productive resources and opportunities over time, are essential.

Some of the early analyses of subsistence/commercial farm household decision-making behavior include: Epstein (1962), Ortiz (1973), Greenwood (1973), Mellor (1966), Edwards (ed. 1968), and Wharton (ed. 1970). More recent studies which have made a relatively successful attempt to integrate farm household decision-making analyses with regional and national policy issues include: Hart (1979), Berry (1975), Greenwood (1976), Hill (1970), Hill (1982), Popkin (1979), Scott (1976), Merrill (1984), and Ewell (1984).

These new approaches have increasingly used historical and political economy concepts in research and analysis. Particularly in Latin America, research methods borrow heavily from neo-Marxist anthropology and economics. Their sensitivity to issues of political power and the framework for comparing different economic systems and their interactions have proven to be helpful in field analysis. Some of the analytical categories employed--"mode of production," "processes of production," "relations of production," "reproduction"--can be profitably utilized even within non-Marxist paradigms (Seddon 1978).

Family Economy Analysis Used in the Tabasco Study

The approach to family economy analysis, which I used in the cacao sector case study in Tabasco, was eclectic. I attempted to collect a very wide range of information on the households and communities studied: historical and current patterns of farm production and access to productive resources; household economic interactions with individuals and institutions outside the household; and changing social relations of production between employers and employees, producers and marketing agents, producers and the state.

A number of assumptions about family economic behavior followed. First, families had multiple objectives, not necessarily farm profits, and they behaved rationally to achieve those objectives. The conditions and constraints on family resources made a qualitative difference in decision-making objectives and strategies.

Second, specific family characteristics may affect production behavior: the demographic stage or "life cycle" of the family; the dependent:worker ratio; and family access to land resources and to community and family support.

Third, basic subsistence needs were defined to include not only minimal food and shelter requirements, but also maintenance of the family's basic farming resources (land, infrastructure, water supplies) and investment in goods, cash, or labor in the community support system which, for example, provided insurance against family disaster (such as a serious illness of the head of household). The minimum acceptable return may be above or below the market wage (Hunt 1979).

Fourth, family resources included the labor of its working members, its land base, cash income, and savings in kind. These could be applied to agricultural activity, on-farm nonagricultural activities, and off-farm agricultural or nonagricultural employment.

Fifth, for families in which land and capital resources were scarce, family labor use was the most significant variable input in the family's strategy. The ratio of producers to consumers in the family determined both the labor resources available and the level of production required for family subsistence.

Finally, farm families with different types of resources and at different demographic stages would produce, market, hire labor, and hire out as workers differently.*

In addition to describing on-farm production patterns, I sought to identify questions of historical and current access to productive resources, and the social relations among the household groups studied.

TABASCO RESEARCH STRATEGY

I attempted to follow the "pyramid" approach in designing a research strategy for the Tabasco study. The key steps are summarized below.

Identify Economic Forces in Tabasco

In order to establish what the most important economic trends in Tabasco state were in the late 1970s, macrodata were used which had been collected by COPRODET and PEMEX showing levels of new investment, growth in production, and new governmental programs. Past analyses of "boom" economies suggested collecting information about inflation rates, wage scale changes, migratory flows, urban population growth, and product and service bottlenecks, much of which was available from COPRODET and other state studies (Havens, Romieux 1966; Kaplan 1980; Nolasco 1979; Pearson 1970; Quintero 1970; Tanzer 1969; Tugwell 1975).

---

*Pertinent literature includes Berry (1975), Chayanov (1966), Greenwood (1978), Greenwood (1973), Hill (1970), Martínez and Rendón (1978), Ortiz (1973), Popkin (1979), and Seddon (ed. 1978).

This data base was supplemented by a large number of open-ended interviews with major Tabascan businesses, unions, PEMEX, government officials, and agricultural leaders. These provided a sense of which groups were losing and gaining from the boom, and what the major state economic and political issues were. Most of the resulting analysis is presented in Chapters 2 and 3.

Choose the Cacao Sector for In-depth Analysis

In order to pursue an in-depth analysis of the changes in the agricultural economy, including production, marketing, and employment analyses, I decided to use a commodity approach. Cacao was chosen partly because it was the most important agricultural sector in the state. It had also been one of the most successful sectors during the 1970s, despite its vulnerability as an export crop to exchange rates, and its highly labor-intensive production system.

Identify the Regional Structure of the Cacao Economy and Petroleum Impacts

I collected and made a preliminary analysis of the major state and national documents available about the history and current organization of the cacao economy in Tabasco, including production systems, technology, marketing, prices, producer organization, and regional variations of these factors.

Interviews with twenty key individuals in Tabasco's cacao community--UNPC officials, leading producers, Cacao Plan tecnicos, BANRURAL credit officers, chocolate industry management--established that labor market disruptions and changes were the principal effects of the oil boom on cacao farming.

A second stage of field research to explore regional variation in cacao economy conditions--farm labor use, wage rates, farm size and technology, cacao politics, employment, and migration--involved 48 interviews with cacao farmers on rancherías and ejidos throughout the state. I also had access to nine cacao community studies made by the Universidad Autónoma de Chapingo at Puyacatengo ("Resúmenes" 1979).

Hypothesize Cacao Farm Family Sub-Groups

Marginal analysis relating wage rate changes directly to farmer migration and agricultural production did not explain the preliminary results of the regional survey. It suggested that in some areas large numbers of cacao family members were indeed flocking to petroleum jobs; but in other areas just the reverse was happening--farmers who used to work off-farm were returning to farm work exclusively. While some farmers were producing unprecedented quantities of cacao (state production levels were at an all-time high), others had nearly ceased all cacao production save maintenance.

The potential typologies that were most commonly used in studies of production response included those based on land tenure, local producer association, and farm size. However, the above contradictions were not elucidated by grouping cacao farmers in any of these ways (see Scherr 1983b, Appendix B charts). Rather, labor seemed to be the critical link between the cacao and petroleum sectors. It was hypothesized that family labor characteristics were the most significant factor for differentiating family household behavior, particularly the land:labor ratio and the role of hired labor.

Test Subgroups Differences

The next step was to test this hypothesis of farmer differentiation, to assure that the subgroups did indeed represent groups with substantially different behavior. The 1978 Tabasco Cacao Census (CCT) (SARH 1978) and the 1980 Sociodemographic Survey of Tabasco (SST) Cacao Subsample data (COPRODET 1980b) were utilized for this purpose. The Cacao Census included data on basic farm and farmer characteristics, cacao production and technology use on nearly 11,000 cacao farms. The SST Cacao Subsample was a more in-depth study of family employment and migration for 408 households, of whom 315 possessed cacao farms. Notes on these studies and my own regional and farm surveys may be found in Appendix B.

Tabasco Cacao Census cross-tabulations showed few significant differences in variables such as tenure or cacao association, whereas among family economy-derived subgroups, a number of variables differed markedly. The SST showed even more marked differences in basic demographic and employment characteristics among the different farmer types. This information was summarized in Chapter 3; more detailed analyses may be found in Scherr (1983b, Appendixes).

Micro-Analysis to Explain Aggregate Behavior

In order to ensure that my field data could be properly interpreted at the state level, I interviewed intensively in three small communities which had quite different economic conditions: an old cacao-growing area in Cunduacán in the heart of the oil producing region; a newer, more marginal cacao zone in Teapa away from most oil disruptions and in the process of transition to a predominantly ranching economy; and a collective ejido with substantial cacao production in the Chontalpa. I collected information on community-level historical and economic conditions which might affect local household decision making and experience. I also collected local data on cacao input costs and marketing information to compare and contrast with that in other parts of the state.

Once it was established that the subgroups did seem to have significantly different characteristics, a small number of farm households representing each of the major types of producers were chosen for detailed farm management and family employment analysis. The purpose of this part of the study was to cross-check explanations of decision making from the more general survey with hard management data and to explore other aspects of decision making which could only be derived from more sustained research. Ejido C-29 was chosen to represent the collective farm producers. Eight farm households were chosen to represent the three different "types" of producer families.

Interview Techniques. Data collection for the collective farm had two parts: personal and group interviews, and use of archives. History, problems, labor use, and ejidal organization were collected through interviewing. The archives contained information on cacao and general production costs, payments of wages and profits, use of hired labor, ejidal income and investments, and cacao production over time.

For the eight case study households, an elaborate questionnaire was devised for application in four visits. This questionnaire is summarized in Appendix B; the full text can be found in Scherr (1983b, Appendix D). The first visit included an introduction and collection of information on family members, farm lay-out, history, and assets. The second visit covered farm activities, employment by activity, and other employment of family members. The third visit was for collection of farm cost and family income information. The last visit was for checking incongruities and completing an open-ended questionnaire on personal plans, historical and projected activities and responses to changing cacao production conditions, and farmer-community interactions.

With the exception of most cacao production data, which was corroborated with receipts from the Cacao Association, and generally at the Association offices as well, all information was via recall. Recall appeared to be quite good for specific detail (how much foliar fertilizer was purchased in March), but quite poor for summary type statistics (total fertilizer purchased in 1979); therefore, the questionnaire was organized to elicit detail.

These data were then analyzed for cost structure and employment patterns and compared to Cacao Plan budget prototypes. Profit levels under different definitions of income and cost were computed. Principal constraints to increased cacao production, both technological and in terms of labor, capital, and land assets, were identified. Questions involving these analyses were referred to the farmers a year later, in a return visit to Tabasco (during a particularly bad cacao season), at which time comparative information on the next year's production and employment were collected.

When integrated and cross-checked with the cacao producer survey, these questionnaires established the systems of cacao production which were used on different types of farms, expected levels and composition of farm expenditures, and the types of responses that might be expected from changes in macroeconomic variables.

Document Subgroup Activities at Macrolevel

The CCT provided overall production information, which showed the importance, in 1978, of each of the cacao household types. Unfortunately, there were no time series data of cacao production for the different groups. The scarcity and unreliability of historical production data, both at the farm level and the Cacao Association level, precluded such analysis during a one and a half year field study. So, rather than concentrate on absolute production levels, this study analyzed the direction and rate of change of economic activity--and the reasons behind those changes--via the various reconnaissance and community-level surveys and case studies. A large random sample survey, stratified by farmer type, could have been used to get a clearer idea of the distribution of key variables among cacao producers which were not covered in the CCT and SST, but for my purposes the effort involved in carrying out the additional study would not have been justified by the results.

The SST did provide considerable information on migration and employment changes over time for different types of households. Differential migration rates for different age groups were established for the farmer types, which corresponded well with the collective histories (in terms of land settlement, and inheritance and farm employment opportunities during different periods) gathered from interviews.

Employment changes between critical economic periods in the state's history: 1965, 1971, 1977, 1978, and 1979 documented the movement of heads of households and of the EAP of cacao families between agricultural wage labor, cacao and other farming, and nonagricultural activities. These supported the overall picture of employment changes revealed in key informant and case study interviews. Together, the CCT, SST, regional surveys, and farm studies from the data base for Chapters 4 through 7.

Analyze Global Cacao Production and Employment Response to Changing Economic Conditions

The next stage of analysis was to examine the dynamic effects on the state economy of decision making by farmers of the different types studied. Chapter 8 documents and analyzes the evolution of the rural labor market in Tabasco as a result of changes in both the cacao and petroleum economies and the global impact on labor supply and demand, land use, production and migration, of the aggregate decisions of the different groups of producers.

A periodic sample survey of producers stratified by household type could be developed from this analysis, which would provide reliable data for technical assistance programs and reliable coefficients for use in estimation of state employment and migration levels, farm production, farm input demand, and sectoral income.

Determine Replicable Factors in Tabascan Farm Response

The final stage of analysis was to determine the extent to which factors in Tabascan agricultural response to oil syndrome distortions were representative of petroleum-producing economies elsewhere, and the extent to which agricultural policies used there might be expected to be useful in other petroleum-producing economies. Chapter 9 summarizes the key variables that are likely to affect agricultural response and the potential role of different types of producers in a petroleum economy.

Appendix B

Notes on Data Sources

Apart from official data sources from COPRODET, SARH, and other governmental organizations on aggregate economic information, and local-level studies described in Chapter 2, there were four major data sets from which most of my analysis was derived. These were the Cacao Census of Tabasco, the Sociodemographic Survey of Tabasco, my own regional cacao producer survey, and my in depth questionnaire for eight farm case studies. Their characteristics, principal problems, analytical methods, and my use of their statistics are described below.

CACAO CENSUS OF TABASCO

SARH of Tabasco carried out an ambitious census of the cacao farmers of Tabasco in 1978. This was the first attempt in the history of the state (apart from the very limited information in the decennial agricultural census) to collect systematic data on farmers. Of an estimated 14,000 cacao producers in Tabasco in 1978, nearly 11,000 were interviewed.

Although I was unable to contact the individuals originally responsible for the Census, I was given to understand that a large group of students, technical workers, and professionals carried out the Census within a several-week period. For many variables, they directly questioned the farmers; for others, they were required to inspect the plantations personally to determine their condition and likely yield. Yield data are calculated from plantation age, size, and condition rather than harvest records.

Its key questions included: property tenure, total area in farm and total area planted to cacao; total number of cacao trees, age distribution of cacao trees, farmer ownership of spraying equipment, use of pest control, and type of insecticide used; use of disease control and type of fungicide; use of fertilizer and type; condition of cacao and shade trees; need for farm drainage; farmer desire to plant more trees and number he wished to plant; cacao association membership; pruning of cacao trees; farmer use of credit, land subject to credit, and lending institution; number of farm dependents. See Scherr (1983b), Appendix Table B-1 for the complete list of questions.

Initial tabulation was simply by variable, at the state and municipal levels. In 1980, I collaborated with Ing. José Luis Santos of the Rainfed Agriculture District II Data Processing Center in Cárdenas, to extend the analysis. We introduced cross-tabulations by tenure, farm size, area in cacao, and association membership, to examine the effects which these had on use of technology, condition of plantation, and yields.

There were a number of problems with the Census, largely ones of interpretation, since there were few explanatory documents with the computer files or results. There was no way to indicate on the form the use of several simultaneous pesticide or fertilization techniques. There was no way to differentiate between the many types of pruning. The definition of persons "dependent" on the farmer was unclear; we took it to be those living in the household, including any resident workers.

Cacao Census Analysis of Farmer Differences

The Cacao Census data was used to test differences in technology and production between farmers grouped by municipal location, land tenure, cacao association, farm size, and area in cacao. As simple explanatory variables, none of these was borne out by the evidence from the 1978 Cacao Census. Data showing the relationship between the variables, and cacao yields and technology use can be found in Appendix Tables B-2 through B-7 in Scherr (1983b).

There were indeed some differences between ejidal and private cacao production. Ejidal structure affected credit procedures, government relations, and access to technical assistance. In many areas ejidal settlements were more recent than private properties and hence showed different tree age composition, and often poorer access to transportation. But there were no consistent differences in input use. In Comalcalco, ejidatarios had better pest and disease control; in Cárdenas, private cacaoteros were better.

UNPC association membership analysis showed that differences in yields between association members were greater than differences between Associations. Some technological differences between municipios were apparent in 1978. Notable were high pruning rates in Cárdenas and Cunduacán; high pest control in Cunduacán and Teapa; high disease control in Cunduacán, Jalpa, and Cárdenas; and high use of credit in Teapa and Huimanguillo. Paraíso, despite much lower rates of chemical input use, had much higher average production in that year. Higher average yields were also found in Cunduacán.

Although the evidence suggested that farm size did affect cacao production in an important way in terms of land use, it did not appear to have a direct relationship with cacao production and yields. Farms with under a hectare planted to cacao had the highest consistent yields per hectare in Cunduacán, Comalcalco, and Cárdenas. In Cunduacán yields declined fairly consistently with size; in Comalcalco somewhat less so. In Cárdenas, yield per hectare declined with size 0-4 hectares, then increased to 10 hectares, then decreased again. Yield per tree for all municipios seemed to follow no clear trend at all as to size. Total farm size seemed to affect use of inputs, such as equipment, pest control, and shade conditions.

Cacao plantation size was positively correlated with ownership of sprayers, percentage using fertilizer, and less so with pruning. On farms with over 20 hectares, there was commonly a decline in percentage using much of the technology. Farms with under a hectare in cacao tended to have lower levels of chemical pest control. Best shade conditions tended to be among farmers with 4-10 hectares in cacao. Cacao conditions were erratic, though tending to be better at the middle size groups.

Overall, analyses of farm production based on these variables were inconclusive or inconsistent.

Analysis of Family Economy Typology

The data was also analyzed in two different ways in terms of the typology explained in the text. In the first analysis, farms were divided into groups based on *area in cacao*, with farms under two hectares in cacao designated as minifundistas; those with 2.1-5.0 as family-labor-based farmers; those with over 5.0 hectares as hired-labor-based farms.

In the second analysis, farms were divided according to a combined measure of cacao area and *total farm size*. This alternative prevented the grouping together of a one-hectare cacao farm with a 50-hectare ranch which had one hectare of cacao. However, it created a different problem in that a ten-hectare farm with two hectares of cacao and eight of bananas was grouped together with a farm having two hectares of cacao and eight of swampland.

The division of farm data according to typology, by combining total farm area and percentage in cacao, was as follows:

No. of hectares	*Mini-fundia*	*No. of hectares*	*Family Farm*	*No. of hectares*	*Hired-Labor-Based*
0-1	All	3-4	61-100%	7.5-10	61-100%
1-2	All	4-5	61-100%	10-20	81-100%
2-3	0-60%	5-7.5	All	20-50	All
3-4	0-60%	7.5-10	All	50+	All
4-5	0-40%	10-20	All		

The collective group was left out of this analysis because it was difficult for the computer to retrieve from the system, although fairly easy to pick out manually. Almost all the collective farmers would be included in the minifundista grouping above, since their family holdings were generally two hectares.

Important differences between farm types have been reported in the text.

SOCIODEMOGRAPHIC SURVEY OF TABASCO

The Sociodemographic Survey of the State of Tabasco was developed as the result of an agreement between the Tabasco State Government (COPRODET) and the National Population Council (CONAPO) in the summer of 1979. The purpose of the project was to collect primary data on employment, migration and demographic patterns in Tabasco, in light of the petroleum boom of the past decade. The survey instrument was designed during fall 1979.

The Sample

The sample was designed to include approximately two percent of households in the state, with each representing fifty existing households. The population was divided into two universes: (1) private households and (2) individuals living in hotels and pensiones. Separate questionnaires were administered. All localities with over 5,000 inhabitants and all municipios were self-representative for most variables.

In order to increase accuracy, the population was divided into "strata"; a random sample was selected from each strata. Urban localities were stratified by number of inhabitants, and internally in the largest cities by community income; rural localities were stratified with respect to predominant economic production regime (banana, cacao, petroleum, cattle, et al.).

The urban sample frame was derived by counting city blocks (stratifying them in larger cities) and choosing without replacement a certain number. All houses on selected blocks were counted and numbered and then selected randomly. The rural sample frame was based on Census Block Information from the national cartography "Geostatistic Regions." One "region" represented 100 households. One such region was chosen for sampling in each economic production regime.

After selection, an exact counting and listing of all homes in the area was undertaken. Then, individual households were chosen from these lists, using a table of random numbers. The number chosen in selected regions was based on a two percent sample of the population in the production regime. When no informant could be found after three visits, or the family had moved away, a brief questionnaire was applied to a neighbor to find out certain characteristics of the family head, and why the family was not at home.

The hotel-pension population was sampled by stratifying establishments by price and type, then choosing a sample randomly. In each selected establishment, a listing was made of individuals who were "habitual" residents there, and from this list, the final sample was randomly selected. This same questionnaire was used for individuals lodging in a family home, but unrelated. Only four such questionnaires were registered as belonging to cacao producers; none among hotel residents.

The Questionnaire

The final questionnaire was composed of the following parts:

I. *Basic Information on the Family Households*
Age, sex, marital status, second residence, temporary migration status, education.

II. *Migration Information*
Birthplace, location and locality size of last two places of residence.

III. *Economic Activities*
Activities of non-EAP individuals over eight years old,

primary and secondary occupations, months occupied, position, branch, income, location of work, type of employer.

IV. *Historical Occupation and Residence*
Whether in work force in 1959, 1965, 1971, 1977, 1978, 1979; primary occupation, position, branch, place of residence in those years.

V. *Agricultural Employment*
Characteristics of family members' agricultural occupations past and present; land ownership present or past; present use of land owned; tenure; land size; primary and secondary products on land owned.

VI. *Coverage of Social Security and Health Care*

VII. *Food Consumption*

VIII. *Emigration and Family Ties*
All members of the household who left in past ten years; reason for leaving, basic characteristics, occupation past and present, present economic ties with family.

IX. *Characteristics of Housing*
Walls, floor, roof, electricity, fuel for cooking, water, sewage, number of rooms, tenancy and rent, state of housing, other uses of housing, number of occupants and renters.

X. *Reproduction and Contraception* (for women over 15 years)

XI. *Means of Communication and Attitudes Toward Present and Future*

This questionnaire took between twenty minutes and two hours to complete, depending upon the number of family members, with an average of 45 minutes. The definition of "household" used was that group of individuals who habitually slept under the same roof, shared food costs, helped with maintenance costs of the house and/or depended economically on the family. It included family members who were temporarily living elsewhere but during 1979 had formed part of the household. In total, over 4,000 households were surveyed.

Analysis of the Cacao Subsample

This survey provided an unusual opportunity to collect and analyze family-based data for cacao farmers. As a consultant for the survey, I was able to presort completed questionnaires for separate analysis. Any questionnaire with mention of cacao as farm product, present occupation, or past occupation was collected. The 408 families with present or historical ties to cacao farming included 126 minifundia, 81 family-labor-based farms, 67 hired-labor-based farms, 33 collective farm households, and 93 landless households.

Due to time and assistance limitations, only the following variables were recorded: locality, municipio, number of members in household, number of members in potential EAP, and number of emigrants. For individuals, information on relation to head of household; age; migratory status; most recent migration; primary and secondary occupations; position, branch, employer and place of occupation; primary occupation in 1978, 1977, 1971, and 1965; and farmer status were collected. Farm information included tenure, land size, principal and secondary land uses.

These data were tabulated to provide state-wide statistics; then the families were divided into farmer types: minifundista, family farm, hired-labor-based farm, collective farm, nonfarm. This division was based on the author's judgment, according to criteria of farm size and farm product mix, given that the total area planted to cacao was not available. SPSS frequency and cross-tabulation analyses were undertaken.

Appendix Table C-1 in Scherr (1983b) shows the basic characteristics of the cacao-producing population in the SST. The original questionnaires asked respondents to state their occupations in their own words. In reviewing the responses, I devised a system of categories which were relevant to this study. Some of the thesis Appendix tables in Scherr (1983b) list the entire breakdown of occupations. The larger categories were:

1. Cacao farming--referring to individuals who grew cacao with or without other crops.
2. Other farming--referring to individuals who farmed or ranched, but did not grow cacao.
3. Agricultural work--referring to individuals who did farm or ranch work for other people.
4. Skilled trades--referring to individuals whose manual work required special training or apprenticeship, such as mechanics, welding, masonry, carpentry.
5. Unskilled work--referring to individuals involved in manual work which did not require special training, such as hauling cargo, domestic work.
6. Trade and commerce--referring to individuals whose activities were in the tertiary sector; included professionals as well as service provid-

ers like beauticians, hotel clerks, and secretaries.

Biases and Problems

Problems in interpreting this survey data stemmed from the following: (1) the Survey was not designed to capture cacao information specifically; (2) there was a good deal of conceptual confusion about terminology; and (3) there was consistent misapplication of several aspects of the questionnaire.

Survey Sample Selected. While the selected sample was representative for the state at large, it does not exactly represent the cacao producer. In a few cases, families were intended to be double-counted. In many cases, the rural areas supposedly intended to represent cacao farmers, were in fact inhabited by many other types, and vice versa. Approximately two percent of the producing families were interviewed. Due to problems in sampling, informant rejection, and probably unsorted questionnaires, producers in Paraíso, Jalpa, and Teapa municipios were under-represented.

A number of families whose head was a cacao worker, rather than producer, escaped the sorting process. The sample was too small to give effective total rates of migration of cacaoteros, even at the municipal level, since emigration has not been a "common" event.

Survey Design. The survey was designed only to capture primary and secondary farm products. Because of this, producers whose tertiary activity was cacao were not captured.

For detailed analysis of labor needs and use it was essential to know how many trees of cacao the farmer had, which we did not know from the SST. Also, my experience was that upon being asked "how much land do you have?" most farmers referred only to the residential or nearby plot. Hence, multiple plot information is probably lacking, and total farm land underestimated.

The historical data on employment was insufficient for more than a general analysis of changes in labor use. Because only five specific years were included, there was an underestimation of the number of people who had at some time worked for PEMEX, for example. Because no information on secondary employment was requested and since field data suggest that most off-farm employment by cacao producers was during short periods of a year otherwise dedicated to cacao, the role of construction and petroleum employment in the family economic strategy was greatly *under*estimated.

The final major design problem was related to part-time family employment. The survey separated questions on economic "employment" and "noneconomic activity." The enumerators were instructed to mark as economically active anyone who worked as much as the equivalent of one month per year. Hence the housewife who also sold chickens all year or the student who on weekends helped his father with cacao were supposed to be considered EAP. Their other role--which was really the principal activity (housewife, student)--was then only mentioned under a section, "During the time you did not work, what did you do?" which did not allow analysis of which activity was the predominant one. It would have been more useful to count "student" or "housewife" directly as an activity (either principal or secondary), and for later analyses simply drop them out as indicated.

Problems of Survey Implementation. Initial instruction to surveyors about the details of collecting employment data were minimal, which led to undercollection of data in the earliest completed questionnaire. Secondary occupation was often not asked in the first few days of the enumeration. Luckily, the majority of the cacao producers were interviewed well into the survey, when the quality of interviewing was better.

The other variable which was underestimated was that of part-time family work. As just discussed, instructions were to count as EAP all family members who worked at least one month. Comparison of separate field data with Survey data shows that in many cases, family members who worked during the cacao harvest were not counted. This was probably because the work was only a few days per month, and although summing to over a month per year, was not perceived as doing so by either informant or surveyor.

The last main source of error in the survey was confusion on the part of enumerators on some terminology. In "Branch of Economic Activity," types of construction were frequently confused (petroleum, housing, urban infrastructure, roads, rural works). Codes on 'for whom one worked' were frequently confused with 'branch of economic activity,' leading to errors as to whether the individual in construction worked for PEMEX, a construction company, the government, or a private firm.

REGIONAL CACAO PRODUCER SURVEY

During the preliminary stage of research, I visited all of the major cacao regions of Tabasco, and interviewed UNPC Association officials for eight local associations, and leaders of other local political and commercial organizations, to discuss conditions in the local economy. I also reviewed detailed community production studies for nine cacao-producing rancherías in the state.

Key questions in the open-ended questionnaire included: local wage rates and conditions of employment; sources of cacao workers; extent of petroleum-related employment; petroleum-related economic problems for cacao farmers; local historical conditions of cacao production; labor demands for cacao production on farms of different size and production intensity; and local adoption of new cacao technologies. I also asked informants to identify which groups in their local area had seen their economic position improve or deteriorate during the 1970s.

Simultaneously, I chose a number of regions, representing the major cacao production systems (cacao and maize; cacao and livestock; cacao and coconut; cacao alone; cacao and bananas), and selected a non-random sample of 48 cacao producers for a brief questionnaire (one half to one hour) on budget issues. The regions were central Comalcalco; Río Séco; eastern Cunduacán; northern Huimanguillo; a Chontalpa ejído; and central Teapa. The questionnaire elicited information on family farm holdings; nonfarm economic interests and employment of family members; wage rates and conditions of employment in cacao; economic changes in cacao and other farm production which had taken place during the 1970s; participation with the Cacao Plan; and the nature and extent of farm labor constraints.

This survey was used to develop the initial hypotheses about cacao farm response to the petroleum boom in Tabasco and to develop the relevant typology of farm producers.

TABASCO CACAO FARM QUESTIONNAIRE

The farm questionnaire was designed using many concepts from other questionnaires, including Puyacatengo community study questionnaires ("Resúmenes" 1979), a CIES (Centro de Investigaciones Ecológicos del Sureste) questionnaire, livestock production analyses from Professor Robert McDowell of Cornell University, farm management analyses from Professor George Conneman of Cornell University, cacao input variables from Cacao Plan documents and the Cacao Census, demographic variables inspired by the SST, and by numerous studies on migration and employment worldwide.

The questionnaire in its entirety may be found in Scherr (1983b, Appendix B). The focus was on seeing how and why farmers performed particular labors on their cacao farms; when they decided to work off-farm and for what reasons; and how they decided to use hired labor. The cost questions were intended both to ensure that all mojor activities had been accounted for, and to check whether the farmers' claims that certain inputs or labor use could or could not be afforded were valid.

There was often a broad difference between what farmers perceived their economic and financial status to be, and what the figures revealed. In almost all cases, when these contradictions were discussed with the farmers, important new factors were revealed. It should be noted here that such problems of computing income make rural income figures from the SST so suspect as to be entirely useless.

The questionnaire was quite long, with each of the four or five sessions lasting one to two hours. However, after they were completed and analyzed, I could find no section whose deletion would not have seriously affected my results. Often this was not because the specific question was so important, but because its answer revealed that some other central question had not been thoroughly or correctly answered. The instrument was designed for considerable cross-checking, and in this seemed quite successful. The same questionnaire, slightly revised, could be used for developing accurate production analyses (farm management studies or linear programming models) for a large sample of cacao farmers, stratified by type, to get more specific data on global input use and production response rates.

Farmer compliance and patience with the extended interviews was remarkable. Families were provided only with a copy of their financial summaries as a return; some of the families were offered, and a few of them accepted, a token M$100 (US$4.50) as well. I am grateful indeed.

Appendix D in Scherr (1983b) provides a discussion of lessons for other researchers stemming from design and analysis of this questionnaire. The most important points were the following:

- One should take care to collect management data from *all* farm plots.
- Such a questionnaire--basically exploratory--should be carried out as early as possible in the research; possibly leaving time for a large-sample random sample survey, stratified by farm family type, on key production elasticity and labor use variables.
- More information should be collected on individual, as well as household expenditures, to determine which expenditures were covered by whose income, and how incomes were distributed within the nuclear and extended family.
- A way to collect employment information directly and privately from older children should be devised; many seemed reluctant to discuss certain issues in front of their parents.

Appendix C

1970 Agricultural Labor Force Estimates For Tabasco

According to the Population Census of 1970, there were 116,147 workers in agriculture in Tabasco, or 59 percent of the EAP. Of these, 58,222 were registered as ejidatarios or propietarios. An additional 12,542 family workers were registered as working principally in agriculture. The Population Census also listed 42,261 jornaleros and 3,122 permanent laborers, indicating that at least 40 percent of the agricultural population was landless in 1970.

One finds a number of discrepancies when comparing this Census information with the Agricultural Census of the following year. The latter registered 163,858 workers in total, or 47,000 more than the Population Census. The bulk of the difference was in the category "producers and family workers," 50,000 more of whom were registered in the Agricultural Census.

Furthermore, the Agricultural Census data on number of workers hired in the harvest period was 41 percent higher than the figures discussed above. The seasonal difference in hired help was slight--only 11 percent. But the seasonal difference between permanent and harvest level "family workers" was 60 percent. This clearly illustrates the critical nature of agricultural labor by family members who do not consider themselves to be principally involved in the agricultural sector.

Glossary

acahual Term referring to unutilized scrubland.

afectación The process by which PEMEX, or another government agency, establishes the right of eminent domain over private or ejidal property.

al tercio An arrangement in which the producer agrees to give a third of his maize harvest in payment for the right to plant on another's land.

anticipo The payment made to a producer upon delivery of cacao to the Association on the basis of the price estimated at the beginning of the season. If the final average price is higher, the difference is distributed later as a "remanente."

BANAMEX (Banco Nacional de Mexico) Private Mexican bank.

BANRURAL (Banco de Crédito Rural, S.A.) The Mexican national bank for rural credit.

cacaotero A farmer who produces cacao.

cacao variety There are three major types of cacao: "criollo," a high-quality variety native to Tabasco; "forastero," a higher-yield but lower-quality variety; and "Ceylán," a hybrid of the other two.

campesino Spanish term for "peasant;" literally, man of the countryside.

campesinado From campesino, referring to the entire peasant population.

CEPAL (Comisión Económica para América Latina) Economic Commission for Latin America (ECLA).

CIES (Centro de Investigación Ecológico del Sureste) The Center for Ecological Research in the Southeast.

CCT (Censo de Cacao de Tabasco) Census of Tabascan cacao producers undertaken by SARH in 1978.

CNIA (Comisión Nacional de la Industria Azucarera) National Commission for the Sugar Industry.

coa A hand-produced wooden implement used in cacao production.

collective farm household A family which belongs to a collective ejido, and farms on both the collective and a private plot.

compadrazgo The institution of interfamily ties based on a relation of "compadres" or co-godparenthood.

comunidad A type of village-based Mexican landholding based on historic indigenous communities. Members are "comuneros."

CONADECA (Comisión Nacional del Cacao) National Cacao Commission, formed in 1974, principally to deal with international marketing. In Chiapas State, CONADECA is chief local marketing agent and in charge of technical programs as well.
CONAPO (Consejo Nacional de Población) National Population Council, located in Mexico City.
CONASUPO (Comisión Nacional de Subsistencias Populares) National Commission for Basic Commodities, the national marketing board for basic foods.
constructora A private construction company, usually large-scale. Many government projects are contracted out to these enterprises.
contratista Individual who contracts with an employer to complete a particular job, for which he independently hires the necessary workers and supervises their work; a private contractor.
convite Traditional institution of communal sowing in the rancherías of Tabasco.
COPLADE (Comité de Planeación para el Desarrollo) Tabasco State Development Planning Commission (formerly COPRODET).
COPLAMAR (Comisión de Planeación para las Areas Marginades) Planning Commission for Marginal Areas.
COPRODET (Comité Promotor de Desarrollo del Estado de Tabasco) Committee to Promote Tabasco State Development.
CROC (Confederación Regional de Obreros de la Construcción) Regional Federation of Construction Workers.
CTM (Confederación de Trabajadores Mexicanos) Mexican Workers' Union, the umbrella organization for Mexican labor unions.

demographic cycle The life-cycle of an individual or family, which changes according to changing demographic characteristics like marriage and raising children, with consequent changes in work and life-style patterns.
demographic stage A stage within the demographic cycle, e.g., the period of early marriage before children, when the new couple may remain with in-laws.
dependency ratio Ratio within the family between the number of "dependents" (nonworking spouses, invalids, students, small children, retired workers) and the number of active workers.
de planta Type of worker who is employed full-time on a contract.

ejido The principal unit of the Mexican agrarian reform; a grant of land usufruct rights to a group of "ejidatarios," either for individual or group farming.
encomienda A charter to collect tribute from a given Indian community, granted by the King of Spain to individuals

in the New World. Many encomenderos acquired rights to labor from the resident indigenous population. Some of the larger encomiendas covered hundreds of thousands of hectares and held thousands of workers.

enganchador An individual who charges a company needing workers a payment for each qualified worker which he finds.

expropriación Expropriation; referring to the takeover of private lands by PEMEX or some other government agency. In return, farmers receive indemnification for the worth of their land (in the case of private property owners) and standing crops and improvements.

family economy A branch of economic analysis which focuses on family economic decision-making strategies as both producers and consumers.

family-labor-based (FLB) farm household Type of farm household with a sufficient land base to support a family without substantial off-farm income, whose farm can be operated primarily with family labor.

fermentation Process by which chemical composition of cacao is changed to release chocolate color and flavor; generally done by piling beans in boxes over a period of days, triggering heat and natural fermentation.

FIRA (Fideicomiso Instituido en Relación a la Agricultura) An organization of the central Banco de Mexico which funnels capital from various international banks into long-term investments in the agricultural sector.

hectare Unit of metric measurement equivalent to 2.47 acres.

hired-labor-based (HLB) farm households Farm households with more land than can be cared for solely with family labor, which thus depend on hired help.

INCATABSA (Industria de Cacao de Tabasco, S.A.) Tabascan Cacao Industry. A group of cacao processing facilities owned by the cacao producer organization.

jornalero A wage worker who finds employment by the day, or "jornal," principally in agriculture. "Jornalear" means to participate in such work.

League of Agrarian Communities Long-established organization which became active in Tabasco in the 1970s to protect political and economic rights of campesinos. In 1980 it had 3,000 members.

libre Type of worker in industry or agriculture who works without a fixed contract.

man-day A unit by which work is measured; a man-day usually corresponds to about 5 to 7 hours of farm work. The "family man-day" refers to a man-day worked by a farm family member.

man-month A month of 25 man-day equivalents.

manga Hand-made implement used in Tabasco for cacao farming.

mano Traditional unit of land measurement for milpa.

mestizo An individual whose ancestry is a "mix" of Spaniard and indigenous.

milpa A subsistence maize plot in Mexico; generally a shifting cultivation system.

minifundia farm household Type of farm household whose land base is insufficient to support a family, and thus is dependent on income from off-farm employment. A "minifundia" refers to such a smallholding; "minifundista" to such a farmer.

municipio Political unit in Mexico roughly equivalent to a county.

Pacto Ribereño Organization of campesinos (with 7,500 members in 1980) providing legal help in defending rights and payment of indemnification by PEMEX in Tabasco.

PEMEX (Petroleos Mexicanos, S.A.) Mexican Petroleum Company, the nationalized oil monopoly of Mexico.

peso Monetary unit of Mexico. Between 1952 and 1976, it was valued at US$.08. With subsequent devaluations, its average worth in dollars fell to US$.045 in 1976, US$.021 in 1982, and US$.007 in 1983.

PIDER (Programa Integrado de Desarrollo Rural) Integrated Rural Development Program.

PRODERITH (Programa de Desarrollo de los Recursos del Trópico Húmedo) Commission responsible to SARH, working since the early 1970s to investigate resources of the humid tropics and develop ecologically, socially, and economically sound programs to exploit those resources.

proletarista Term used in the debate on agrarian development in Mexico, referring to those who are convinced that the process of rural proletarianization is leading to the demise of the campesino economy.

propiedad Mexican tenure designation for private property ownership; the maximum size is limited by law. "Propietario" refers to the owner of such a private property.

pruning Important activity in cacao production to shape trees properly. The principal types of pruning in Tabasco cacao farming are: "desmadre" (removal of main stems), "desmamone" (removal of young shoots), "deschupone" (removal of suckers), and "poda" (early pruning of formation).

ranchería A small, unincorporated rural community.

recampesinista Term used in the debate on agrarian development in Mexico, referring to those who are convinced that the campesino economy is an integral part of the dominant capitalist economic system, and will continue to play a viable role for the foreseeable future.

remanente A payment made to cacaoteros for their cacao after marketing has been completed, usually based on profits from international sales (see *anticipo*).

rial Traditional unit of land measurement for milpa.

SAG (Secretaría de Agricultura y Ganadería) Ministry of Agriculture and Ranching (until 1976).

SAM (Sistema Alimentario Mexicano) Mexican Food System, established at the end of the Lopez Portillo administration to coordinate efforts to increase agricultural production by low-income farmers, and food consumption by low-income consumers.

SARH (Secretaría de Agricultura y Recursos Hidráulicos) Ministry of Agriculture and Hydraulic Resources (after 1976).

sexenio A period of six years; in Mexico, a reference to the term in office of political office-holders.

shade trees Principal varieties of shade trees for cacao plantations in Tabasco are moté (*Erythrina* sp.), cocoite (*Gliricidia* sp.), samán and chipilín.

SPP (Secretaría de Programación y Presupuesto) Ministry of Planning and Budget (after 1981).

SRH (Secretaría de Recursos Hidráulicos) Ministry of Hydraulic Resources (until 1976).

SST (Sociodemographic Survey of Tabasco) The "Encuesta Sociodemográfica del Estado de Tabasco," undertaken in 1979 as a joint effort between the Tabasco government and CONAPO.

STPRM (Sindicato de Trabajadores Petroleros de la República Mexicana) The Union of Petroleum Workers of the Mexican Republic.

técnico An individual who has been trained in some technical field.

tequio Traditional system for communal work projects, such as school- and road-building.

transitorios Type of industrial worker hired through a temporary contract.

UNPC (Unión Nacional de Productores de Cacao) National Cacao Producers' Association.

Bibliography

GENERAL

"A Survey of Nigeria." *The Economist* (January 26, 1982).

Alianza Popular No. 3. Volume of presentations at state political conference, Villahermosa, Tabasco, October 1974.

"Alimentación, Crisis Agrícola y Economía Campesina." *Comercio Exterior* 28, No. 6 (June 1979), Mexico, D.F. Unsigned editorial.

Allub, Leopoldo. "Heterogeneidad estructural, desigualdad social y privación relativa en regiones petroleras." *Revista Mexicana de Sociología* 45, No. 1 (1983).

--- and Marco A. Michel. "Migración y estructura ocupacional en una región petrolera." *Revista Mexicana de Sociología* 44, No. 1 (January 1982).

--- and Marco A. Michel. *Industria Petrolera y Cambio Regional en México: El Caso de Tabasco.* Mexico, D.F.: Cuadernos del Centro de Investigación para la Integracion Social, 1980.

Álvarez Luna, Eduardo. "Perspectivas de las Zonas Tropicales de México Como Productor de Alimentos." Unpublished speech, July 10, 1978.

Amuzegar, Jahangir. "Oil Wealth: A Very Mixed Blessing." *Foreign Affairs* (Spring 1982), pp. 814-35.

Antonio Ortiz, Ing. Martín. *El Cacao en México.* Villahermosa, Tabasco: Unión Nacional de Productores de Cacao, 1964.

"Armed Peasants Halt Mexican Oil Production." *Modesto Bee*, January 12, 1980.

BANAMEX. *Review of the Economic Situation of Mexico* 60, No. 705 (August 1984).

Banos Ramirez, Othón. "Los campesinos tabasqueñas a la hora del petroleo: el caso de los cacaoteros." *Yucatán: Historia y Economía*, Año 5, No. 28 (November-December 1981):26-58.

Barbosa-Ramirez, A. René. *La Ganadería Privada y Ejidal: Un Estudio en Tabasco*. Mexico: Centro de Investigaciones, 1974.

Barkin, David. "El Mercado de Trabajo en la Chontalpa." Comisión Económica para América Latina. Naciones Unidos: Consejo Económico y Social, November 1976.

--- and Blanca Suárez. *El Fin de la Auto-Suficiencia Alimentaria*. Mexico: Editorial Nueva Imagen, 1982.

--- et al. *Evaluación del Plan Chontalpa*. Mexico: Centro de Eco-Desarrollo, 1975.

Barlett, Peggy. "Microdynamics of Debt, Drought and Default." Paper presented at American Agricultural Economics Association Annual Meeting. Ithaca, New York, August 5-8, 1984.

---, ed. *Agricultural Decision Making*. New York: Academic Press, 1980.

Benito, Carlos A. "Farmers' Response to Rural Development Projects in Puebla." In Roger D. Norton and Leopoldo Solis, eds., *The Book of CHAC: Programming Studies for Mexican Agriculture*. Baltimore, Md.: Johns Hopkins University Press (for the World Bank), 1983.

Bermúdez, Antonio J. *Le Política Petrolera Mexicana*. Mexico: Joaquín Mortiz, 1976.

---. *The Mexican National Petroleum Industry*. Stanford, Calif.: Institute of Hispanic American and Luso: Brazilian Studies, Stanford University Press, 1963.

Berry, Sara S. "Households, Decision Making, and Rural Development: Do We Need to Know More?" Development Discussion Paper No. 167, Harvard Institute for International Development. Cambridge, Mass.: Harvard University, May 1984.

---. *Oil and the Disappearing Peasantry: Accumulation, Differentiation, and Underdevelopment in Western Nigeria*. Boston: Boston University African Studies Center, 1982.

---. *Cocoa, Custom and Socioeconomic Change in Rural Western Nigeria*. Oxford: Clarendon Press, 1975.

Bienen, Henry. "Oil Revenues and Policy Choice in Nigeria." Washington, D.C.: World Bank Staff Working Paper No. 592, 1983.

Binswanger, Hans P. and Mark R. Rosensweig. *Contractual Arrangements, Employment and Wages in Rural Labor Markets: A Critical Review*. New York: Agricultural Development Council, 1981.

Blandford, David. "The World Cocoa Market--Structure, Dynamic and Control." Unpublished manuscript, Cornell University, Department of Agricultural Economics, 1980.

Bolís Villanueva, Eduardo. *La Economía del Estado de Tabasco*. Mexico: Sistemas Bancos de Comercio, March 1976.

Bracho Roblín, Ing. Emilio; Ing. Guillermo Vázques Rodríguez, and Ing. Marco Antonio Wing Martínez. *Situación del Cacao en México*. Mexico, D.F.: Fondo de Garantía y Fomento para la Agricultura Ganadería y Avicultura, 1972.

Braudeau, Jean. *El Cacao*. Barcelona: Editorial Blume, 1970.

Braun, Armelle. "Rethinking Agricultural Development: Algeria's Adjustment Process." *Ceres #88* 15, No. 4 (July-August 1982):39-45.

Brecher, Richard and Ian Parker. *Cocoa, Employment and Capital in the Ghanaian Economy*. New Haven: Yale University Press, 1974.

Bromley, Ray and Chris Grey, eds. *Casual Work and Poverty in Third World Cities*. Chichester: John Wiley and Sons, 1979.

Casco Montoya, Rosario. *Los Planes de Desarrollo del Trópico: El Caso de Balancán-Tenosique, Tabasco*. Mexico: Centro de Ecodesarrollo, 1980.

---. *Manejo del Agua en un Ecosistema Tropical: El Caso de la Chontalpa*. Mexico: Centro de Ecodesarrollo, 1979.

CEPAL (ECLA). *Economía Campesina y Agricultura Empresarial*. Mexico, D.F.: Siglo XXI, 1982.

Chayanov, A.V. In Daniel Thorner, Basile Kerblay, and R.E.F. Smith, *The Theory of Peasant Economy*. Homewood, Ill.: American Economics Association and Richard D. Irwin, 1966.

Cleave, John H. *African Farmers: Labor Use in the Development of Smallholder Agriculture*. New York: Praeger, 1974.

Consulta Técnica Sobre Métodos y Técnicas de Extensión Agrícola para el Cacao. La Alianza de los Paises Productores de Cacao (The Cocoa, Chocolate and Confectionery Alliance), el Gobierno de la República Federativa del Brazil (CEPLAC), and Food and Agriculture Organization. Itabuna/Ilheus, Bahia, Brasil, May 1977.

Conway, Patrick and Alan Gelb. "Oil Windfalls in a Controlled Economy: A 'Fix-Price' Equilibrium Analysis of Algeria." Discussion paper, Development Research Department, World Bank. Washington, D.C., 1984.

Corden, W.M. "Booming Sector and Dutch Disease Economics: Survey and Consolidation." *Oxford Economic Papers* (forthcoming).

--- and J.P. Neary. "Booming Sector and Deindustrialization in a Small Open Economy." *The Economic Journal* 92 (December 1982):825-48.

Croll, Donald O. "Production Decline Arrested." *Petroleum Economist* (January 1984):25-27.

de Janvry, Alain and Carmen Diana Deere. "A Theoretical Framework for the Empirical Analysis of Peasants." CUDARE Working Paper No. 60. Berkeley: University of California, 1978.

de Navarrete, Ifigénia M., ed. *Bienestar Campesina y Desarrollo Económico*. Navarrete. Mexico: Fondo de Cultura Económica, 1971.

"Diagnóstico de la Situación Actual y Definición del Tipo de Integración Vertical." Estudio de Gran Visión para el Programa de Desarrollo de la Chontalpa, Tabasco. Cárdenas, Tabasco, February 1976.

"Diez Años Después: La Nueva Polémica Agraria." *Nexos* 6, No. 71 (Mexico, D.F., November 1983):25-37.

Eckstein, Salomón. *El Ejido Colectivo en Mexico.* Mexico: Fondo de Cultura Económica, 1966.

Edwards, Edgar O., ed. *Employment in Developing Nations.* New York: Columbia University Press, 1968.

Ehrensaft, Philip, Pierre La Rameé, Ray D. Bollman, and Frederick H. Buttel. "The Microdynamics of Farm Structural Change in North America: The Canadian Experience and Canada-U.S.A. Comparisons." Paper prepared for presentation at annual meeting of the American Agricultural Economics Association, August 1984, at Cornell University. Mimeographed.

El Mallakh, Ragaei, Oystein Noreng, and Barry W. Poulson. *Petroleum and Economic Development.* Lexington, Mass.: Lexington Books, 1984.

"Energy Problems of the Third World." *Petroleum Economist* (February 1984):501.

Epstein, Scarlet. *Economic Development and Social Change in South India.* Manchester: Manchester University Press, 1962.

Esteva, Gustavo. "Los campesinos existen." *Nexos* (September 1983).

---. "Y si Los Campesinos Existen?" *Comercio Exterior* 28, No. 6 (Mexico, D.F., June 1978).

Ewell, Peter. "The Intensification of Peasant Agriculture in Yucatan." Cornell International Agricultural Economics Study, A.E. Research 84-4 (May 1984), Cornell University.

--- and T.T. Poleman. *Uxpanapa: Resettlement and Agricultural Development in the Mexico Tropics.* New York: Pergamon Press, 1980.

FAO (Food and Agriculture Organization). *Production Yearbook.* New York: United Nations, 1982a.

---. *Trade Yearbook.* New York: United Nations, 1982b.

Falcon de Gyves, Zaida. *Análisis de los Mapas de Distribución de la Población del Estado de Tabasco.* Mexico: Instituto de Geografía, Universidad Nacional Autónoma de México, 1965.

Foro Internacional 72. Vol. 18, No. 4 (April-June 1978).

Fox, Edward. *History in Geographic Perspective.* New York: Norton, 1971.

Frauvergue, S. *Changements dans l'Isthme de Tehuantepec: la Raffinerie de Salina Cruz.* Paris: Institute des Haute Études de L'Amerique Latine, Doc. de Recherche No. 3, 1980.

Frazier, Steve. "Mexico's Boom Damped by Oil Glut." *The Wall Street Journal*, August 13, 1981, p. 27.

Gelb, Alan H. "Adjustment to Windfall Gains: A Comparative Analysis of Oil Exporting Countries." Draft, 1984, Development Research Department. Washington, D.C., World Bank.

---. "Capital-Importing Oil Exporters: Adjustment Issues and Policy Choices." Washington, D.C.: World Bank Staff Working Paper No. 475, 1981.

Gerhardt, Peter. *The Southeast Frontier of New Spain.* Princeton, N.J.: Princeton University Press, 1979.

Gillis, Malcolm, Dwight H. Perkins, Michael Roemer, and Donald R. Snodgrass. *Economics of Development.* New York: W. W. Norton and Company, 1983.

Gladwin, Christina and Robert Zabala. "Microdynamics of Contraction Decisions: A Cognitive Approach to Structural Change." Paper presented at the American Agricultural Economics Association meeting, August 1984, at Cornell University.

Glassburner, Bruce. "Oil Public Policy and Economic Performance: Indonesia in the 1970's." Unpublished paper, 1984.

Gliessman, S.R., R.E. Garcia, and A.M. Amador. "The Ecological Basis for the Application of Traditional Agricultural Technology in the Management of Tropical Agro-Ecosystems." Unpublished manuscript, Colegio Superior de Agricultural Tropical, Cárdenas, Tabasco, 1979.

Gómez Oliver, Luís. "Crisis Agrícola, Crisis de los Campesinos." *Comercio Exterior* 28, No. 6 (Mexico, D.F., June 1978).

Gordon, David. "Clouds of People." Survey in *The Economist*, April 22, 1978.

Gray, B.A. *The Nigerian Petroleum Industry: A Guide*. Washington, D.C.: Library of Congress, Maktaba Afrikana Series, 1978.

Grayson, George W. "Mexico's Opportunity: The Oil Boom." *Foreign Policy on Latin America, 1970-80*. Boulder, Colo.: Westview Press, 1983.

Green, R.H. and S.H. Hymer. "Cocoa in the Gold Coast: A Study in the Relations Between African Farmers and Agricultural Experts." *Journal of Economic History* 26, No. 3 (1966):299-319.

Greenwood, Davydd. "Community-Level Research, Local-Regional-Governmental Interactions and Development Planning: A Strategy for Baseline Studies." Paper for USAID, May 1978, Cornell University Department of Anthropology.

---. "The Political Economy of Peasant Family Farming: Some Anthropological Perspectives on Rationality and Adaptation." Rural Development Occasional Paper No. 2, Cornell University Center for International Studies, 1973.

Gunder Frank, Andre. *Capitalism and Underdevelopment in Latin America*. New York: Monthly Review Press, 1969.

Hall, Lana and Turner Price. "Price Policies and the SAM: A Wheat-Maize Comparison." Cornell University, Department of Agricultural Economics, 1981.

Hart, Gillian. "Strategies of Labor Allocation in Rural Javanese Households." Ph.D. dissertation, Cornell University, 1978.

Havens, A. Eugene and Michel Romieux. *Barrancabermeja: Conflictos Sociales en Torno a Un Centro Petrolero*. Ediciones Tercer Mundo y Facultad de Sociología, Universidad Nacional de Colombia, Bogotá, 1966.

Helleiner, George. "Smallholder Decisionmaking: Tropical African Evidence." In Lloyd Reynolds, ed., *Agriculture in Development Theory*, pp. 27-52. New Haven and London: Yale University Press, 1975.

Hernández Marqués, Baldemar. "Análisis del Costo de la Vida en el Municipio de Cárdenas." Unpublished manuscript, Colegio Superior de Agricultura Tropical, 1979, Cárdenas, Tabasco.

--- y Manuel R. Villa Issa. "Análisis del Mercado de Trabajo en el Plan Chontalpa." Unpublished manuscript, Colegio Superior de Agricultura Tropical, 1978, Cardenas, Tabasco.

Hill, Polly. *Dry Grain Farming Families.* Cambridge, England: Cambridge University Press, 1982.

---. *The Migrant Cocoa Farmers of Southern Ghana: A Study in Rural Capitalism.* Cambridge: University Press, 1970.

Hunt, Diana. "Chayanov's Model of Peasant Household Resource Allocation." *Journal of Peasant Studies* 6, No. 3 (April 1979).

IMF (International Monetary Fund). *Financial Statistics.* Washington, D.C.: IMF, 1984.

Johnston, Bruce. "A Comparative Analysis: Agricultural Development in the U.S., Mexico and Taiwan." In Bruce Johnston, Cassio Luiselli, Roger Norton, and Celso Cartas Contreras, eds., *U.S.-Mexico Relations: Agriculture and Rural Development.* Stanford, Calif.: Stanford University Press (forthcoming).

--- and Peter Kilby. *Agriculture and Structural Transformation.* London: Oxford University Press, 1975.

Just, Richard E. and David Zilberman. "Risk Aversion, Technology Choice, and Equity Effects of Agricultural Policy." Giannini Foundation of Agricultural Economics Paper, December 1984, University of California at Berkeley.

Kahl, Joseph. *Modernization, Exploitation and Dependency in Latin America - Germani, González Casanova and Cardoso.* New Brunswick, N.J.: Transaction Books, 1976.

Kaplan, Marcos. "Petroleo y Desarrollo: El Impacto Interno." *Foro Internacional.* Colegio de Mexico, July-September 1980.

Katouzian, M.A.H. "Oil vs. Agriculture: A Case Study of Dual Resource Depletion in Iran." *Journal of Peasant Studies* 5, No. 3 (April 1978).

Kolawole, Michale Iyiola. "An Economic Study of Traction on Farm Systems in Eastern Upper Volta." Ph.D. dissertation, Cornell University, 1972.

Kraft, Joseph. "The Mexican Oil Puzzle." *The New Yorker* (October 1979):150-81.

LeClair, Edward E., Jr. and Harold K. Schneider, eds. *Economic Anthropology: Readings in Theory and Analysis*. New York: Holt, Rinehart and Winston, 1968.

Link, John E. "Mexico Aims at Self-Sufficiency in Basic Foods, Reduced Imports." *Foreign Agriculture* (January 1981).

Lipton, Michael. *Why Poor People Stay Poor: Urban Bias in World Development*. Cambridge, Mass.: Harvard University Press, 1977.

Looney, Robert E. *Economic Origins of the Iranian Revolution*. New York, Pergamon Press, 1977.

López Portillo, José. "Tercer Informe del President de la República de México." *El Heraldo de México*, Sunday, September 2, 1979, Año XIV, No. 4974.

Lustig, Nora, ed. *Panorama y Perspectivas de la Economía Mexicana*. Memoria del Coloquio Sobre Economía Mexicana, El Colegio de Mexico, March 1979.

Maddock, Rodney and Ian McLean. "Supply-Side Shocks: The Case of Australian Gold." *Journal of Economic History* 44, No. 4 (December 1984).

Madjd, Mohammad Gholi. *Land Reform and Agricultural Policy in Iran, 1962-78*. New York: Cornell University, Department of Agricultural Economics, 1983.

Maier, Elizabeth. *Chinampa Tropical - Una Primera Evaluación*. Mexico: Maier, Centro de Ecodesarrollo, 1979.

Mamalakis, Markos and Clark Reynolds. *Essays on the Chilean Economy*. Homewood, Ill.: R. D. Irwin, 1965.

Martínez, Marielle P. L. and Teresa Rendón. "Fuerza de Trabajo y Reproducción Campesina." *Comercio Exterior* 28, No. 6 (Mexico, D.F., June 1978).

Martínez Najera, Julio. *La Economía del Estado de Chiapas*. Mexico, D.F.: Sistemas Bancos de Comercio, December 1975.

Martz, John D. "Approaches to Agricultural Policy in Venezuela." *Inter-American Economic Affairs* 34, No. 3 (Winter 1980).

Marx, Karl. *Capital*. New York: Vintage Books, 1976.

Matlon, Peter. "Guidelines for Agricultural Surveys: Notes from Nigeria." Paper presented at the Department of Agricultural Economics, Cornell University, 1978.

McCleery, Robert and Clark W. Reynolds. "The Impact of Migration on the U.S. Economy: Selected Issues." Unpublished paper, 1984.

McDowell, R.E. *Improvement of Livestock Production in Warm Climates.* San Francisco: W. H. Freeman and Company, 1972.

--- and P.E. Hildebrand. "Integrated Crop and Animal Production to Optimize Resource Utilization on Small Farms in Developing Countries." Paper presented at Bellagio Study and Conference Center, October 18-23, 1978, Lake Como, Italy. Published by the Rockefeller Foundation.

Mears, Leon A. "Rice and Food Self-Sufficiency in Indonesia." *Bulletin of Indonesian Economic Studies* 20, No. 2 (August 1984):122-38.

Megateli, Abderrahmane. *Investment Policies of National Oil Companies.* New York: Praeger, 1980.

Mellor, John W. *The Economics of Agricultural Development.* Ithaca: Cornell University Press, 1966.

Merrill, Deborah. "The Mixed Subsistence-Commercial Production System in the Peasant Economy of Yucatan, Mexico: An Anthropological Study of Commercial Beekeeping." Ph.D. dissertation, Cornell University, 1984.

---. "The Economic Role of Children in Peasant Societies: Toward an Approach to Research." Department of Agricultural Economics Working Paper, December 1978, Cornell University.

"Mexico: Crisis of Poverty/Crisis of Wealth." *Los Angeles Times*, July 15, 1979.

Meyer Cosío, Lorenzo. "El Auge Petrolero y Las Experiencias Mexicanas Disponibles: Los Problemas del Pasado y La Visión del Future." *Foro Internacional* 18, No. 4 (Colegio de Mexico, April-June, 1978).

Michel, Marco Antonio and Leopoldo Allub. "Petroleo y Cambio Social en el Sureste de Mexico." *Foro Internacional* 18, No. 4 (Colegio de Mexico, April-June, 1978).

Mikesell, Raymond F. "Petroleum Exploration." *Energy Policy* (March 1984):17-18.

Muñoz Ruíz, Humberto. *Mejoramiento del Cultivo del Cacao en Tabasco.* Chapingo, Mexico: Universidad Autónoma de Chapingo, Ingeniería Agrícola-Industria, 1956.

Nankani, Gobind. "Development Problems of Mineral-Exporting Countries." World Bank Staff Working Paper No. 354, 1979, Washington, D.C.

Negrete Salas, María Eugenia. "Petroleo y Desarrollo Regional: El Caso de Tabasco." *Economía Demográfica* 18, No. 1 (Colegio de Mexico, 1984), 57.

Negrón, Marco. "Prediagnóstico Para Un Plan de Desarrollo Integral de la Faja Petrolífera del Orinoco." *Foro Internacional* 18, No. 4 (Colegio de Mexico, April-June 1978).

Nolasco, Margarita. *Ciudades Perdidas de Coatzacoalcos, Minatitlán y Cosoleacaque.* Mexico, D.F.: Centro de Ecodesarrollo, 1979.

Norman, D.W. "Methodology and Problems of Farm Management Investigation: Experience from Northern Nigeria." African Rural Employment Paper No. 8, 1973, Department of Agricultural Economics, Michigan State University.

Norton, Roger D. and M. Leopoldo Solis, eds. *The Book of* Baltimore: The Johns Hopkins University Press, 1983.

Odell, Peter R. "Outlook for the International Oil Market and Options for OPEC." *Energy Policy* (March 1984):5-12.

---. "The Oil Industry in Latin America." In Edith T. Penrose, ed., *The Large International Firm in Development Countries: The International Petroleum Industry.* London: George Allen and Unwin, 1968.

Okorie, Aja and David Blandford. "World Market Trends and Prospects for Cocoa." Cornell International Agriculture Mimeograph 73, September 1979.

Okurume, G.E. *Foreign Trade and the Subsistence Sector in Nigeria: The Impact of Agricultural Exports on Domestic Food Supplies in a Peasant Economy.* New York: Praeger, 1973.

Olayemi, J.K. and S.A. Oni. "Costs and Returns in Peasant Cocoa Production: A Case Study in Western Nigeria." *Bulletin of Rural Economics and Sociology* 6, No. 2 (1971):133-62.

"One-party Misrule." *The Economist*, September 4, 1982, p. 34.

Onoh, J.K. *The Nigerian Oil Economy from Prosperity to Glut*. New York: St. Martin's Press, 1983.

Ooi Jin Bee. *The Petroleum Resources of Indonesia*. Oxford, England: Oxford University Press, 1982.

Ortiz, Sutti. *Uncertainties in Peasant Farming*. London: The Athlone Press, University of London, 1973.

O'Shaughnessy, Hugh. *Oil in Latin America*. London: The Financial Times, 1976.

Oxaal, Ivar, Tony Barnett, and David Booth, eds. *Beyond the Sociology of Development: Economy and Society in Latin America and Africa*. London: Routledge and Kegan Paul, International Library of Sociology, 1975.

Padilla Paez, Hermes. "Algunos Aspectos de la Producción y del Comercio del Cacao en Mexico." Thesis, 1971, Escuela Nacional de Agricultura, Chapingo, Mexico.

Papenek, Gustav. *The Indonesian Economy*. New York: Praeger, 1980.

Pearson, Scott. *Petroleum and the Nigerian Economy*. Stanford: Stanford University Press, 1970.

--- and John Cownie. *Commodity Exports and African Economic Development*. Lexington, Mass.: Lexington Books, 1974.

Penny, David. "Hints for Research Workers in the Social Sciences." Occasional Paper, 1973, Cornell University, Department of Agricultural Economics.

Pfeiffer, Karen. "Effects of Recent Public Policy Changes on Agrarian Markets in Algeria." Paper presented to Society for Economic Anthropology Meeting, April 1984, at Davis, California.

Pirez, Pedro. "Modalidades de desarrollo y politica regional en México, 1960-1980." *Revista Mexicana de Sociología* 45, No. 1 (1983).

Popkin, Samuel. *The Rational Peasant: The Political Economy of Rural Society in Vietnam.* Berkeley: University of California Press, 1979.

Price, Turner and Lana Hall. "Agricultural Development in the Mexican Tropics: Alternatives for the Selva Lacandona Region of Chiapas." A.E. Research 83-4, January 1983, Cornell/International Agricultural Economics Study, Cornell University.

Producción es Revolución. Reforma, Chiapas, Mexico: Municipio de Reforma, 1977.

Pucciarelli, Alfred. "Contradicciones del desarrollo regional polarizado. El papel de la agricultura en la micro-región Lázaro Cárdenas." *Revista Mexicana de Sociología* 45, No. 1 (1983).

Purseglove, J.W. *Tropical Crops: Monocotyledons 1 and 2.* New York: John Wiley and Sons, Halsted Press Division, 1972.

---. *Tropical Crops: Dicotyledons.* London: Longman Group, 1968.

Quinlan, Martin. "Oil Policy under the Generals." *Petroleum Economist* (February 1984):55-57.

Quintero, Rodolfo. *El Petroleo y Nuestra Sociedad.* Caracas: Ediciones de la Biblioteca, Universidad Central de Venezuela, 1970.

Ramirez, Carlos. "El Plan de Desarrollo, Condena de hambre a quienes tienen menos." *Proceso* 344 (June 1983):7-15.

Relations of Production: Marxist Approaches to Economic Anthropology. London: Frank Cass, 1978.

Rendón, Teresa. "Empleo en Mexico." Unpublished paper, 1975, Colegio de Mexico.

"Resúmenes, Estudios de Caso de Comunidades en Tabasco." Unpublished student field research, August 1979, Universidad Autonoma de Chapingo, Puyacatengo.

Reyes Osorio, Sergio, Rodolfo Stavenhagen, Salomón Eckstein, Juan Ballesteros, et al. *Estructura Agraria y Desarrollo Agrícola en México.* Mexico, D.F.: Fondo de Cultura Economica, 1974.

Reynolds, Clark. *The Mexican Economy*. New Haven: Yale University Press, 1970.

Richards, Alan. "The Food Crisis and State Policy in the Middle East." Working paper toward a research proposal, 1983. Mimeographed.

Rippy, Merrill. *Oil and the Mexican Revolution*. Leiden: E.J. Brill, 1972.

Rodríguez Cisneros, Manuel et al. *Características de la Agricultura Mexicana*.

Rodríguez Vallejo, Ing. José. "La Oferta y La Demanda de Máiz en México." Paper presented at the National Symposium, "El Maíz en Mexico: Su Pasado, Presente y Futuro," February 9-11, 1981, at Guadalajara, Jalisco, Mexico.

Roemer, Michael. "Dutch Disease in Developing Countries: Swallowing Bitter Medicine." Development Discussion Paper No. 156, October 1983, Harvard Institute for International Development. Mimeographed.

Rose, Beth. *An Overview of the Indonesian Rice Economy*. Ithaca: Cornell University, Department of Agricultural Economics, 1982.

Roseberry, William. "Peasants, Proletarians, and Politics in Venezuela, 1875-1975." In Robert P. Weller and Duke Guggenheim, *Power and Protest in the Countryside*. Durham, N.C.: Duke Press Policy Studies, 1982.

Rosensweig, Fernando. "Política Agrícola y Generación de Empleo en México." *El Trimestre Económico* 42, No. 4 (October-December 1975), #168.

Salazar-Carrillo, Jorge. *Oil in the Economic Development of Venezuela*. New York: Praeger, 1976.

Salgado Escarpita, Jose. *La Economía del Estado de Veracruz*. Mexico: Sistema Bancos de Comercio, 1976.

Sanders, Thomas. "The Economic Development of Tabasco, Mexico." *Fieldstaff Reports* 5 (No. 8, American Universities Field Staff, 1977).

Sayegh, Kamal S. *Oil and Arab Regional Development*. New York: Praeger Special Studies in International Economy and Development, 1976.

Schatzl, L.H. *Petroleum in Nigeria*. Ibadan: Oxford University Press, Nigeria Institute of Social and Economic Research, 1969.

Scherr, Sara J. "Agriculture and the Oil Syndrome: The Role of Public Policy on Developing Petroleum Economies." Paper submitted to the International Association of Agricultural Economists Meeting, August 26-31, 1985, Malaga, Spain. Mimeographed.

---. "Incorporating Macroeconomic Variables in Anthropology Research." In John Bowen, ed., *Anthropology in the Development Process*. Lanham, Md.: University Press of American and Society for Economic Anthropology (forthcoming).

---. "Agriculture in Developing Petroleum Economies: The Historical Record and Policy Options." Draft, 1984

---. "Resolving the Agriculture-Petroleum Conflict: The Experience of Cacao Smallholders in Mexico." Cornell International Agricultural Economics Study, A.E. Research 83-33, August 19, 1983a, Cornell University.

---. "Resolving the Agriculture-Petroleum Conflict: The Experience of Cacao Smallholders in Mexico." Ph.D. dissertation, Cornell University, 1983b.

---. "Tendencias Demográficas en Tabasco Desde 1970." Report to the Comité Promotor de Desarrollo Socioeconómico del Estado de Tabasco, August 1979, Villahermosa, Tabasco.

---. "Livestock Production in the Mexican Southeast." Unpublished term paper for "Livestock Production in the Tropics," Department of Animal Science, Cornell University, May 1978.

---. "The Mexican Development Strategy: Retrospect and Prospect." Cornell Agricultural Economics Staff Paper #75:28, December 1975, Cornell University.

--- and T.T. Poleman. "Development and Equity in Tropical Mexico: Thirty Years of the Papaloapan Project." Cornell International Agricultural Economics Study, January 1983, Cornell University.

--- and Robin D. Scherr. "Tabasco and the Petroleum Boom: Prospects for State Development." Unpublished paper, May 1981.

Schliephake, Konrad. *Oil and Regional Development.* New York: Praeger, 1977.

Schmink, Marianne. "Household Economic Strategies: Review and Research Agenda." *Latin American Research Review* 19, No. 3 (1984):87-101.

Schneider, Jane and Peter Schneider. *Culture and Political Economy in Western Sicily.* New York: Academic Press, 1976.

Schultz, Theodore W., ed. *Economics of the Family: Marriage, Children and Human Capital.* Chicago: National Bureau of Economic Research with University of Chicago Press, 1974.

Scott, James C. *The Moral Economy of the Peasant: Rebellion and Subsistence in Southeast Asia.* New Haven: Yale University Press, 1976.

Seddon, David, ed. *Relations of Production.* London: Frank Cass, 1978.

Shaner, W.W., P.F. Philipp, and W.R. Schmehl. *Farming Systems Research and Development.* Boulder, Colo.: Westview Press, 1982.

Shanin, Teodor, ed. *Peasants and Peasant Societies.* Marmondsworth, Middlesex: Penguin Books (Penguin Modern Sociology Readings), 1971.

Shiras, Peter. "The Political Economy of Smallholder Agriculture: Current Proposals for the Third World in Historical Perspective." Department of Agricultural Economics, Cornell University, 1978.

Silos, José S., Luz María Bassoco, and Victor M. Lozano. "El Sector Agrícola: Comportamiento y Estrategia de Desarrollo." Comisión Nacional Coordinadora del Sector Agropecuario, Nota Técnia No. 6, September 1975, Mexico, D.F.

Smith, Patrick and Geoffrey Weston. "South Special Report--Nigeria." *South* (February 1984):47-57.

Stavenhagen, Rodolfo. *Las Clases Sociales en las Sociedades Agrarias.* Mexico, D.F.: Siglo Vientiuno Editores, S.A., 1969.

Stavins, Robert. "The Nursing, Planting and Basic After Care of Amazon Cocoa: A Guide for Extension Workers." MANR, Agriculture Division, Kenema, Sierra Leone, August 1972.

Sweezy, Paul M. *The Theory of Capitalist Development: Principles of Marxian Political Economy*. New York: Modern Reader Paperbacks, 1942.

Tanzer, Michael. *The Political Economy of International Oil and the Underdeveloped Countries*. Boston: Beacon Press, 1969.

Taracena, Alfonso. *Cunduacán, La Atenas de Tabasco*. Mexico: Editorial JUS, 1978.

Tello, Carlos. *La Política Económica en México 1970-76*. Mexico: Siglo Veintiuno Editores, S.A., 1979.

Thompson, Roberto G. "Petroleo y Desarrollo en el Area de Reforma al Sureste de Mexico 1970-1979." Centro de Investigaciones Económicos y Sociales, August 1979, Villahermosa, Tabasco.

Timmer, C. Peter. "Macro Prices and Structural Change." *American Journal of Agricultural Economics* (May 1984): 196-201.

---, Walter P. Falcon, and Scott R. Pearson. *Food Policy Analysis*. Baltimore: Johns Hopkins University Press, 1983.

Tirado, Manlio. *El Problema del Petroleo: Tabasco, Chiapas y el Gasoducto*. Mexico: Ediciones Quinto Sol, S.A., 1978.

Toledo, Alejandro, ed. "Petroleo y ecodesarrollo en el sureste de México." Centro de Ecodesarrollo, 1982, Mexico.

Trapido, Paul J. *Venezuela, An Export Market Profile*. Washington, D.C.: U.S. Department of Agriculture, 1984.

Tugwell, Franklin. *The Politics of Oil in Venezuela*. Stanford: Stanford University Press, 1975.

Turrent Fernández, Antonio. "The Constraints for Small Farm Agriculture in Mexico." In Bruce F. Johnston, Cassio Luiselli, Roger D. Norton, Celso Cartas Contreras, eds., *U.S.-Mexico Relations: Agriculture and Rural Development*. Stanford: Stanford University Press (forthcoming).

Ul Haque, Irfan. "The Oil Economy Syndrome: Analysis and Policies." Energy Discussion Paper, 1982. Cavendish Laboratory Energy Research Group, Cambridge, England.

Unión Nacional de Productores de Cacao (UNPC). *Boletines Informativos*. Villahermosa, Tabasco, 1977-83.

Uphoff, Norman T. and Warren F. Ilchman, eds. *The Political Economy of Development*. Berkeley: University of California Press, 1972.

Urquhart, D.H. *Cocoa*. London: Longmans, Green and Company, 1955.

Verduzco, Gustavo. *Campesinos Itinerantes*. Mexico, D.F.: El Colegio de Michoacán, 1982.

Warman, Arturo. *Ensayos sobre el Campesinado en México*. Mexico, D.F.: Editorial Nueva Imagen, 1980.

---. "Frente a la Crisis, Política Agraria o Política Agrícola?" *Comercio Exterior* 7, No. 6 (Mexico, D.F., June 1978).

Warr, Peter. "Exchange Rate Protection in Indonesia." *Bulletin of Indonesian Economic Studies* 20, No. 2 (1984): 53-89.

Watts, Michael. *Silent Violence: Food, Famine and Peasantry in Northern Nigeria*. Berkeley: University of California Press, 1984.

--- and Paul Lubeck. "The Popular Classes and the Oil Boom: A Political Economy of Rural and Urban Poverty." In I.W. Zartman, ed., *The Political Economy of Nigeria*. New York: Praeger, 1983.

West, R.C., N.P. Psuty, and B.G. Thom. *Las Tierras Bajas de Tabasco en el Sureste de México*. Villahermosa: Gobierno del Estado de Tabasco, 1976.

Wharton, Clifton R., Jr., ed. *Subsistence Agriculture and Economic Development*. Chicago: Aldine, 1970.

Wilson Salinas, Patricia. "El Auge Petrolero y la Planeación Regional en México." *Revista Interamerican de Planificación* 17, No. 66 (June 1983):149-63.

World Bank. *World Development Report, 1984*. New York: Oxford University Press, 1984.

---. *Energy in the Developing Countries.* Washington, D.C.: The World Bank, 1980a (August).

---. *Indonesia: Employment and Income Distribution in Indonesia.* Washington, D.C.: A World Bank Country Study, 1980b.

Yang, W.Y. *Methods of Farm Management Investigations.* Agricultural Development Paper no. 64. Rome: Food and Agriculture Organization, 1976.

Yates, Frank. *Sampling Methods for Censuses and Surveys.* London: Charles Griffin, 1949.

Yates, Paul Lamartine. *Mexico's Agricultural Dilemma.* Tucson: University of Arizona Press, 1981.

Yúñez Naude, Antonio. "Política Petrolera y Perspectivas del Desarrollo de la Economía Mexicana, Un Ensayo Exploratorio." *Foro Internacional* 18, No. 4 (Colegio de Mexico, April-June 1978).

Zapata, Francisco et al. *Las Truchas.* Mexico, D.F. El Colegio de Mexico, 1978.

Zavala de Cossio, Ma. Eugenia. "Industria petrolera y cambio sociodemográfico: Algunos efectos sobre la población del crecimiento del complejo industrial Coatzacoalcos-Minatitlán-Cozaleacoque." Presented at El Coloquio Franco-Mexicano, September 1981, Mexico, D.F.

GOVERNMENT DOCUMENTS -- MEXICO

CONADECA (Comisión Nacional de Cacao). "Alianza para la Producción: Evaluación 1979." Mexico, D.F. 1980.

---. "Alianza para la Producción de Cacao, 1978." Secretaría de Comercio, Mexico, D.F., 1979.

---. "Desarrollo del Sector Cacaotero 1977." Mexico, D.F. 1978.

---. "Nueva Política para el Desarrollo del Sector Cacaotera, 1976." Mexico, D.F. 1977.

---. "Indicadores Económicos del Cacao 1975." Mexico, D.F. 1976. Volumes for 1977, 1978, and 1979 were also published in 1978, 1979, and 1980, respectively.

---. "1974: Un Año al Servicio del Productor." Mexico, D.F., 1975.

CNEP (Comisión Nacional para la Eradicación del Paludismo). "Estimaciones de la Población en Tabasco." 1970-79.

CNIA (Comisión Nacional de Investigaciones Agrícolas). "El Cultivo de la Caña de Azucar en Tabasco." Serie Recomendaciones, Circular No. 3, Instituto Mexicano para la Producción de Azucar. Mexico, D.F., October 1975.

COPRODET (Comité Promotor de Desarrollo del Estado de Tabasco). "Analisis de las Inversiones Públicas 1959-1979." Villahermosa, Tabasco, 1980a.

---. "Índice de Precios." Villahermosa, Tabasco, 1980b.

---. "Actualización y Depuración de la Matríz de Insumo-Producto del Estado de Tabasco, Informe Final." Villahermosa, Tabasco, 1979a.

---. "Evaluación de los Impactos Socioecológicos de PEMEX en la Zona de Explotación del Cretácico, Ejido de Ceiba Jahuactal, Tabasco." *Plan Estatal de Desarrollo.* Villahermosa, Tabasco, 1979b.

---. "Plan Estatal de Desarrollo, Evaluación y Actualización." *Plan Estatal de Desarrollo.* Villahermosa, Tabasco, 1979c.

---. "Plan Sectorial: Agropecuario y Forestal." *Plan Estatal de Desarrollo* (Segunda Versión). Villahermosa, Tabasco, December 1979d.

---. "Propuesta de Regionalización para el Estado de Tabasco 1979." *Plan Estatal de Desarrollo.* Villahermosa, Tabasco, 1979e.

--- and CONAPO (Consejo Nacional de Población). "Diagnóstico y Política Demográfica del Estado de Tabasco." Villahermosa, Tabasco, May 1981.

--- and CONAPO. "Encuesta Sociodemográfica del Estado de Tabasco." Mexico, D.F., January 1980.

CONAPO (Consejo Nacional de Población). *Política Demográfica Regional: Población Tabasco.* CNP: Mexico, D.F., 1979.

Ejido C-29, Plan Chontalpa. "Libro de Aportación, 1979." Tabasco, Ejido C-29 (original records), 1979.

---. "Librode Contaduría, Cacao, 1979." Tabasco, Ejido C-29 (original records). 1979.

FIRA (Fondo Instituido en Relación con la Agricultura). *Características y Finalidades de los Fondos Instituidos en Relación con la Agricultura en el Banco de Mexico, S.A.* Mexico, D.F.: July 1978.

Gobierno del Estado de Tabasco. "Manual de Estadísticas Básicas del Estado de Tabasco." Villahermosa, Tabasco. February 1980a.

---. "Tabasco en Cifras." Secretaría de Promoción Económica, Villahermosa, Tabasco. 1980b.

---. "Análisis del Costo de la Vida en el Estado de Tabasco." Villahermosa, Tabasco. 1979a.

---. "Convenio Industrial." Programa de la Alianza para la Producción, Villahermosa, Tabasco. 1979b.

Gobierno del Estado de Veracruz. "Poza Rica." Promoción Industrial. 1967.

INEGI/COPLADET (Instituto Nacional de Estadística, Geografía e Informática and Comité de Planeación para el Desarrollo del Estado de Tabasco). *Anuario Estadístico de Tabasco 1984*, Volumes I and II. Mexico, D.F.: INEGI, 1984.

Legislación de la Secretaría de Hacienda y Crédito Público. "Incentivos Fiscales." In *Diario Oficial*, March 6. Mexico, D.F.: 1979.

Legislación para Producir Mas Alimentos (Sere: Ediciones de la Cámara de Diputados 1981). Mexico, D.F.: Cia. Periodística del Sol de México, 1981.

Ley de Asociaciones Ganaderas y Su Reglamento (Unión Ganadera Regional de Tabasco). Villahermosa, Tabasco: 1977.

Ley Federal de Reforma Agraria. Mexico, D.F.: Libro-Mex Editores, 1976.

PEMEX (Petroleros Mexicanos, S.A.). *La Industria Petrolera en México*. Mexico, D.F.: Con la Secretaría de Programación y Presupuesto, October 1980.

---. *Petróleos Mexicanos 1938-1978*. Mexico, D.F.: Secretaría de Educación Pública, 1979.

---. "Informe del Director General de PEMEX." Ing. Jorge Díaz Serrano, March 1978a, Mexico, D.F.

---. *Memoria de Labores 1977.* Mexico, D.F.: March 1978b.

---. *Poza Rica: Apuntes para su Historia.* Mexico, D.F.: 1977.

PIDER (Programa Integrado de Desarrollo Rural). *Monografía del Estado de Tabasco.* Mexico, D.F.: 1976.

Promotora del Complejo Agroindustrial de la Chontalpa. "Promotora del Complejo Agro-Industrial de la Chontalpa: Programa Para 1979." Cárdenas, Tabasco. 1979.

---. "Estudio Social y de Organización Campesina." Documento No. 2, Estudio de Gran Visión para el Programa de Desarrollo Agroindustrial de la Chontalpa, May 1976, Cárdenas, Tabasco.

SARH (Secretaría de Agricultura y Recursos Hidraúlicos). "El Desarrollo Agropecuario de Mexico: Pasado y Perspectivas." Mexico, D.F. 1982.

---. *Econotécnia Agrícola: Consumos Aparentes de Productos Agropecuarios para los Años de 1925 al 1980*, Vol. I, No. 9 (September 1981a), Mexico, D.F.

---. *Econotécnia Agrícola: Situación de la Agricultura Nacional en 1980 y Algunas Consideraciones Sobre el Mercado Internacional*, Vol. V, No. 1 (January 1981b), Mexico, D.F.

---. "Plan de Incremento de la Producción de Cacao en el Estado de Tabasco." Villahermosa, Tabasco. 1980.

---. "Censo de Cacao de Tabasco." Cárdenas, Tabasco. 1978a.

---. *Econotécnia Agrícola: El Cultivo de Cacao*, Vol. II, No. 12 (December 1978b), Mexico, D.F.

---, CESPA (Centro de Estudios de Planeación Agropecuario). SARH-United Nations/CEPAL Project. Mexico, D.F.: SARH, 1982.

---, Comisión de la Grijalva. "Estudio Socio-Económico Definitivo, SARH Proyecto: Desarrollo Agropecuario Integral de la Isla, Centro, Tabasco." Cárdenas, Tabasco. April 1979.

---, DGEA (Dirección General de Economía Agrícola). "Serie Estadística: Producción Agrícola Nacional 1970-1979." Datos Definitivos, Mexico, D.F. 1980.

---, Distrito de Temporal No. 8. "Producción de máiz de invierno 1971-1977." Villahermosa, Tabasco. 1978.

---, INIP (Instituto Nacional de Investigaciones Pecuarias). "Programa del Instituto Nacional de Investigaciones Pecuarias para el Plan Nacional de apoyo a la Agricultura de Temporal." Secretaría de Agricultura y Recursos Hidraúlicos, Comisión de la Grijalva, Dirección de Estudios y Proyectos. 1980.

---, INIP. "Rendimientos y Costos de Produccion de Cultivos Tropicales." Cárdenas, Tabasco. August 1979.

---, PRODERITH. "Programa de Desarrollo Rural Integrado del Trópico Húmedo: Resúmen General, PRODERITH en Tabasco." Villahermosa, Tabasco. September 1978.

---, Residencia de Promoción y Programas. "El Proyecto Piloto 'Zapotal' en Cifras." Villahermosa, Tabasco. 1979.

---, Sub-Programa de Producción y Extensión Agrícola. "El cultivo de Cacao." Folleto Técnico No. 5. Villahermosa, Tabasco. December 1979.

SIC (Secretaría de Industria y Comercio), Dirección General de Estadística. *Séptimo, Octavo, Noveno y Décimo Censo Nacional de Población 1950, 1960, 1970, 1980.* Mexico, D.F.: 1953, 1963, 1972, and 1982.

---. *Censo Agrícola, Ganadera y Ejidal de 1950, 1960 y 1970.* Mexico, D.F.: 1954, 1965, and 1975.

Serfin, Estudios Económicos. *El Sistema Alimentario Mexicano: La Estrategia Definida para el Sector Agropecuario* No. 26:21. Mexico, D.F., 1980.

SAM (Sistema Alimentario Mexicano). *Estrategia de Comercialización y Distribución de Alimentos Básicos e Insumos Productivos.* Mexico, D.F.: December 1980a.

---. *Estrategia de Comunicación Social.* Mexico, D.F.: December 1980b.

---. *Medidas Operativas Agropecuarias y Pequeras, Estrategia de Comercialización, Transformación, Distribución y Con-*

sumo de los Productos de la Canasta Básica Recomendable. Mexico, D.F.: May 1980c.

---. *Primer Planteamiento de Metas de Consumo y Estrategia de Producción de Alimentos Básicos para 1980-1982.* Mexico, D.F.: March 1980d.

---. *Notas Analíticas y Lineamientos Metodológicos para el Proyecto Sistema Alimentario Mexicano.* Mexico, D.F.: August 1979.

Index

About the Author

SARA JEANNETTE SCHERR is currently a Rockefeller Foundation post-doctoral fellow in agriculture. She received her B.A. in Economics from Wellesley College in 1975 and her M.S./Ph.D. in Agricultural Economics from Cornell University in 1983. She has published papers on rainforest exploitation in Costa Rica, agricultural development projects in tropical Mexico, and agricultural policy in oil-exporting countries. She has worked as a consultant for the Tabasco State Government in Mexico, the U.S.-Mexico Project of Stanford University, and has taught development economics at California State University, Stanislaus. This book was written in 1984-85, while Dr. Scherr was a Visiting Scholar at Stanford University's Food Research Institute.